Fuels, Explosives and Dyestuffs

to
Sheila, Anne and Wendy
for their patience and encouragement

The cover photograph shows part of the
hydrodesulphurization unit at Shell's
Pernis Refinery, The Netherlands

'The whole fabric of modern civilization becomes every day more interwoven with the endless ramifications of applied chemistry.'

Leo Baekeland, 1938

Chemistry in Industry

Fuels, Explosives and Dyestuffs

Peter Tooley BSc MSc ARIC MIBiol
Head of Department of Chemistry
St Osyth's College, Clacton

John Murray Albemarle Street London

Titles in this series

High Polymers
Fats, Oils and Waxes
Fuels, Explosives and Dyestuffs
Food and Drugs
Techniques
Handbook of Experiments

© Peter Tooley 1971

All rights reserved.
No part of this publication may be
reproduced, stored in a retrieval system,
or transmitted, in any form or by any means,
electronic, mechanical, photocopying, recording
or otherwise, without the prior permission of
John Murray (Publishers) Ltd,
50 Albemarle Street, London, W1X 4BD

Printed in Great Britain
by William Clowes & Sons, Limited
London, Beccles and Colchester

0 7195 2029 0 boards edition
0 7195 2030 4 limp edition

Preface

The rapid and often revolutionary advances in applied chemistry since World War II have made it increasingly difficult for the hard-pressed teacher to keep abreast of events and bridge the gap between school and industry. In addition the pupil, in daily contact with novel materials forged by the chemists' skill, and keen to discuss the related problems of scientific advance such as pollution and the misuse of knowledge, is unlikely to be satisfied by a narrow academic approach to the subject. In tracing the origin, current development and socio-economic implications of a score or so branches of chemical technology, the present series aims at providing the teacher with relevant background material. Emphasis has been placed upon changing patterns of industry such as the transition from batch to continuous production and other problems of large-scale manufacture. An attempt has also been made to show how the properties of substances are related to their structure and how these properties can be modified by molecular tailoring.

Two complementary volumes, *Techniques* and *Handbook of Experiments*, contain suggestions for related practical work and a survey of modern preparative and analytical techniques.

Although the text has been written primarily as a source-book for science teachers, it should also prove useful to students at universities and colleges as well as senior pupils carrying out project work. Bearing in mind the wide spectrum of the chemical industry covered it is inevitable that certain errors and omissions will have occurred but every effort has been made to give a concise and accurate account of the chosen topics.

Acknowledgments

It is a pleasure to acknowledge the ready help of many colleagues and friends in industry and Government departments who provided valuable advice, statistics and other material. I have also had to draw upon the published works of many authors who are too numerous to mention individually, but to whom my thanks are also due.

I am grateful for the help afforded by the staff of John Murray who have successfully steered the book through the hazards of production. Mention must also be made of the loyal assistance of my senior technician, Miss Sandra Fairbairn, who has not only deciphered and typed the MS but given freely of her time and technical expertise.

Thanks are due to the following who have kindly permitted the reproduction of copyright photographs: cover, Shell Petroleum Co. Ltd; pages 43, 78, Woodhall Duckham Construction Co. Ltd; 51, 54, 64, 116, 149, 189 (*top*), Shell Petroleum Co. Ltd; 76, Humphreys & Glasgow Ltd; 82, 173, 186, 189 (*bottom*), 197, 203, 235, Imperial Chemical Industries Ltd; 107, Chemical Engineering Wilton Ltd; 120, 131, BP Chemicals (UK) Ltd; 133, British Geon Ltd; 183, Crown Copyright.

Thanks are also due to the following for permission to base the diagrams shown on material from their publications: figures 2.3, 2.13, 2.14, 2.15, 2.16, 2.17, 2.18, 2.19, 2.20, 2.28, 2.30, 4.7, Shell Petroleum Co. Ltd; 2.7, National Coal Board; 2.11, Woodhall Duckham Construction Co. Ltd; 2.21, Gas Council; 2.27, *Science Journal*; 2.29, Esso Petroleum Co. Ltd; 3.3, 3.4, 3.5, BP Chemicals (UK) Ltd; 4.3, Imperial Chemical Industries Ltd; tables on pages 6, 30, National Coal Board; 32, Gas Council; 160, BP Chemicals (UK) Ltd.

Contents

1 **Introduction** 1

2 **Fuels** 11
  Solid fuels 21
  Liquid fuels 47
  Natural gas 90

3 **Chemicals from coal and petroleum** 103
  Coal chemicals 103
  Petroleum chemicals 110

4 **Explosives** 161
  High explosives 163
  Special uses of explosives 181

5 **Dyestuffs** 192
  Colour and chemical constitution 206
  The classification of dyes 212
  Dye types 213
  Dyeing synthetic fibres 232

Index 241

Chapter 1
Introduction

> *Mrs B.* There are, however, circumstances which frequently prevent the regular and final decomposition of vegetables. . . . In these cases they are subject to a peculiar change, by which they are converted into a new class of compounds called bitumens . . . they are sometimes of an oily liquid consistence, as the substance called naphtha.
> *Caroline.* These are substances I never heard of before.
>
> *Conversations on Chemistry*, 1828

Energy is one of the fundamental requirements of a community. Just as the human body ceases to function properly and finally dies without adequate supplies of food, so the economy of a country grinds to a halt if its power supplies are inadequate. A fundamental difference between these two energy requirements, however, is that whereas man's appetite for food has a natural limit his appetite for industrial energy is enormous and apparently insatiable. The demand for electricity is doubling each decade, and the world demand for petroleum is expected to have more than doubled between 1956 and 1976. It has been estimated that half the coal used since the beginning of time has been consumed in the last quarter of a century, and half the oil used in the last decade.

Although the world's energy reserves appear to be assured for a long while to come they are haphazardly distributed by nature and the problem of energy production varies considerably from country to country. The problem of transporting 'raw' energy is a major headache and is often resolved by spectacular and massive projects. For example, the Russians aim to build a 1 500 kV direct current line from the Caspian Sea coalfields and a 2·44 m (8 ft) diameter natural gas pipeline from Siberia to W. Russia. An interesting feature of the fuel pattern is that while coal production is steadily falling in Britain, France, and Germany, world production is rising in most other coal producing countries, including the USSR and the USA.

The increase in the demand for energy is certain to continue in the foreseeable future, and it is interesting to see how the pattern of primary energy sources has altered during the last 50 years. Coal has traditionally been a major source of power in Europe since the Industrial Revolution. Indeed coal production probably started in Britain during the twelfth century, following the discovery of 'sea-coles' washed up on the north-east coastline, and reached 7 million tonnes a year by 1750. During the period following World War I, however, the contribution made by coal towards European energy requirements has been shrinking steadily from over 80% in 1920 to 48% in 1967. Meanwhile the share of petroleum, which was first drilled in 1859 and only became of real importance during the early 1920's, has expanded to 32% in the same period. Even more spectacular has been the steady rise in the usage of natural gas consequent upon the recent strikes in Holland, France, Italy, West Germany and under the North Sea. The swing away from coal reflects the rising costs of mining and the advantages of liquid and gaseous fuels, such as their high calorific ratings and convenience of use, storage and transportation.

The advent of nuclear power after World War II introduced a new dimension in energy production. It was even predicted that by the turn of the century atomic fuel would replace traditional fuels.

Sir Robert Robinson, introducing a series of lectures on petroleum chemicals at Manchester College of Science and Technology in 1959, commented: 'If and when we enter the true era of nuclear energy, the chemical industry of carbon compounds will surely leap ahead and provide many of our chief needs, clothing, building materials, and even food, as well as medicinals, cosmetics, dyes, detergents and a host of minor trimmings. In this Utopia the squandering of carbon resources by their use as fuel will be prevented by legislation.'

Indeed one of the attractive features of the use of atomic fuel is that it can only be used for producing energy and does not involve the destruction of valuable raw materials. As Bronowski put it: 'Thorium has been a drug on the market since gas mantles went out of fashion, and no one is going to extract uranium from granite in order to colour porcelain.'

Nevertheless nuclear energy is still the most expensive form of energy and only produces a small but expanding part of world

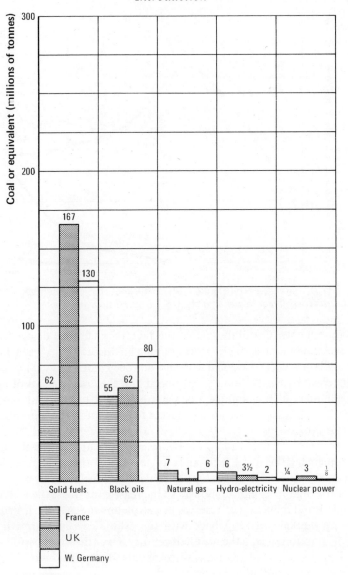

Fig. 1.1   Consumption of primary energy in Europe, 1967

Fuels, Explosives and Dyestuffs

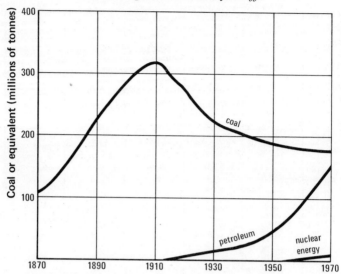

Fig. 1.2 Relative contribution of petroleum, coal and nuclear energy towards energy requirements of the UK (after Hutton–Wilson)

requirements. Certainly one of the factors which has delayed the development of a nuclear programme in the UK has been high installation costs. Coupled with this has been a determined marketing drive by the National Coal Board and the development of new techniques such as fluidized bed combustion. Coal burning power stations now represent a basic market for 40% of the coal produced in Britain.

It seems clear that the probable role of nuclear energy will be to supplement the fossil fuels and not to replace them. The next 50 years may also see a greater use of coal as both fuel and raw material for the organic chemical industry. One reason for this is the increasingly rapid depletion of reserves of petroleum and gas, fresh strikes being unlikely to keep pace with the estimated rapid growth in consumption during the next half century. Over the same period the fall in known coal reserves will probably be less than 10%. This reasoning is probably behind current research being conducted in the USA into the possibilities of commercially exploiting the production of oil from coal by novel routes.

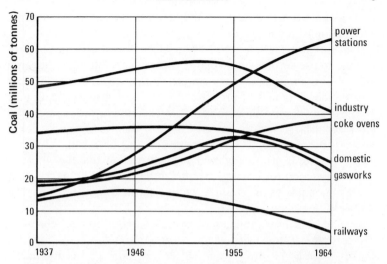

Fig. 1.3 Changing pattern of coal consumption in the UK, 1937–1964

Fuel	Million tonne equivalents of coal	Contribution to total %
Coal and lignite	6 000	33–43
Oil	4 000–6 000	28–33
Natural gas	1 000	6–7
Water power	1 000	6–7
Nuclear energy	2 000–4 000	14–22
Total	14 000–18 000	

Predicted pattern of world energy sources by AD 2000 (after Schumacher)

An interesting example of social legislation was the passing of the Clean Air Act in 1956. This not only made the emission of dark smoke from industrial premises an offence, but gave local authorities power to set up 'smokeless zones' to regulate the emission of smoke from domestic dwellings. The implementation of this Act further diminished the sales of smoky high volatile coals which had already been in decline since 1950. This was due mainly to the increasing use of electrical and diesel traction on the railways, the spread of oil-

and gas-fired central heating, and latterly the use of naphtha in the production of town gas.

This situation provided the National Coal Board with the problem of how to use surplus high volatile coal. As a result much research went into the manufacture of smokeless fuels from low grade coal and the development of processes for the complete gasification of coal without the production of coke. The cost of producing town gas by coal carbonization in vertical retorts had soared and it could be made more economically by steam cracking petroleum naphthas or using natural gas. In any event, estimates show that the whole of the UK demand for town gas, representing about 15% of the total energy demands of the country, will be satisfied by the production of North Sea gas.

One of the disadvantages of using coal for organic chemical production is its low hydrogen to carbon ratio—most coal containing less than 5% hydrogen and 80–90% carbon. This is due to the high proportion of condensed aromatic ring structures present in coal. Petroleum-based fuels on the other hand contain principally saturated aliphatic hydrocarbons and therefore have a much higher proportion of hydrogen to carbon.

	Carbon %	Hydrogen %	Hydrogen/Carbon ratio		Molecular weight
			Gross	Net[a]	
Coal					
Anthracite	over 93	under 4	0·52	0·35	
Bituminous	84–91	5–5·5	0·68	0·65	over 2 000
Lignite	60–75	5·5	1·00	0·25	
Oil					
Fuel oil	86·0	11·8	1·60	1·25	365
Gas-oil	86·2	12·8	1·80	1·80	265
Kerosine	86·4	13·6	1·90	1·90	170
Gasoline	85·3	14·7	2·00	2·00	100

[a] The net hydrogen/carbon ratio involves effective hydrogen after allowing for hydrogen combining and removed with the oxygen, sulphur and nitrogen present in the fuel.

Carbon and hydrogen in coal and oil

Introduction

Processes for converting coal into liquid or gaseous fuels and organic chemicals are therefore designed to increase the hydrogen/carbon ratio from about 1:16 to about 1:8, the latter being close to the hydrogen/carbon ratio found in petroleum, and therefore more suitable for the production of the more highly saturated chain hydrocarbons. For the same reason, although coal tar is an excellent source of cyclic compounds such as benzole, naphthalene and phenol, it is unsuitable for the production of the short chain hydrocarbons which are so much in demand. This would necessitate breaking open the ring structures, a process requiring considerably more energy than that required to crack the long chain hydrocarbons present in petroleum.

For a hundred years organic chemicals remained coal-based before the growth of the petroleum industry. Although manufacture of chemicals from petroleum did not begin on a commercial scale until the early 1920's, this is now the preferred source of hydrocarbon raw materials used in the production of organic chemicals. Over 90% of the organic chemicals currently produced in the USA are already based upon petroleum or 'wet' natural gas. It has been predicted that by 1975 petroleum chemicals will account for half the total output of the US chemical industry, representing some 156 million tonnes a year. This pattern is being repeated in Europe and the output of petroleum chemicals from the West European area reached 50% of the US total in 1968, and is expected to equal it by the 1980's.

Explosives are in effect rapidly burning fuels which are able to exert sudden pressure on their surroundings by the rapid evolution of large volumes of hot gas. A commercial explosive must fulfil a number of requirements. It must be relatively cheap to produce and while remaining stable under normal conditions must explode readily when suitably detonated. To be effective it must also produce a large volume of gas exothermically in order to heat the products of reaction and thus increase their pressure. Solid or liquid explosives are commonly used, not only because of their convenient form, but because of their greater violence due to the increased evolution of hot gas per unit volume.

Although gunpowder was the only explosive used for many centuries, the rise of organic chemistry in the nineteenth century saw

Fuels, Explosives and Dyestuffs

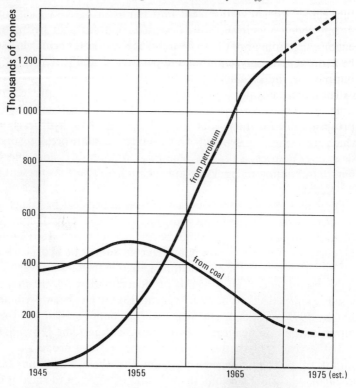

Fig. 1.4 *Production of organic chemicals in the UK*

the development of many new explosives. Most of these were either esters of nitric acid such as nitroglycerine and guncotton, or nitrated coal tar derivatives such as TNT.

Two types of explosive were now emerging. The slower burning propellant explosives which did not reach their peak pressure instantaneously and the newer detonating explosives which could be exploded by shock to give a powerful shattering effect ('brisance'). The propellants, such as cordite, are ideally suited for firing projectiles, such as bullets and shells. Detonating explosives on the other hand are used as bursting charges in bombs, torpedoes, depth-charges and the like. They also have a number of non-military applications, such as quarrying, mining, demolition and

Introduction

tunnelling. High explosives, such as TNT, can be detonated but do not explode on ignition. This is not only a useful safety factor but enables the contents of unexploded bombs and shells to be destroyed by burning. Nitroglycerine not only explodes on detonation but on burning also and is known as a primary explosive.

The development of special explosives for coal mining, and for other commercial purposes such as metal forming and seismic prospecting, reflects current interest in this field.

Commodity	1958	1963	Unit
Blasting powder	4 275	1 250	tonnes
High explosives (a) gelatinous (b) non-gelatinous	60 250	297 600 134 500	tonnes tonnes
Propellants (cordite and other smokeless powders)	1 975	3 600	tonnes
Safety fuse	393 (1 028)	50 (156)	Mm (× million feet)
Electric detonators	143	114	× 1 000
Fireworks (a) amusement (b) signal rockets etc.	2·6 0·9	2·3 1·0	£10⁶
Live ammunition (a) below 30 mm (b) above 30 mm	6·4 1·5	4·7 3·4	£10⁶

Production of explosives in the UK

The rise of coal tar chemistry in the late Victorian era not only led to the discovery of new explosives but also to the manufacture of the first artificial dyestuffs. These interesting chemical substances were almost all obtained from natural sources until the middle of the last century. The discovery of 'Aniline Purple' by Perkin in 1856 revolutionized the dyestuffs industry overnight and paved the way for a host of other synthetic dyes. The production of these early dyestuffs from coal tar marked the genesis of the organic chemical industry and the breaking down of the barriers which existed between pure chemistry and industry. Coupled with the name of Perkin is that of Peter Griess whose work was largely responsible for the development of the azo dyes—about two thousand of which are now available commercially.

The development of man-made fibres has produced many problems for the dye chemist and technologist. The attractive water-repelling characteristics of the synthetics and their small pore size have required the development of new processes and new types of dyestuff to produce satisfactory fast shades of colour. The annual production of synthetic dyestuffs in the UK is currently about 30 000 tonnes.

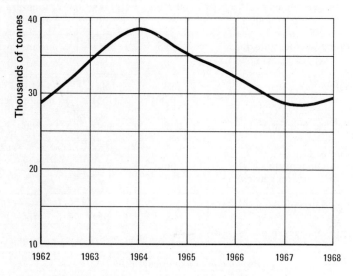

Fig. 1.5   Production of organic dyestuffs in the UK

Interest in organic dyestuffs has inevitably stimulated research into the origin of colour and its relation to molecular structure. The early chromophore theory of Witt has recently been interpreted in terms of current knowledge concerning the electronic configurations of organic molecules.

Chapter 2
Fuels

A large part of the world's energy requirements have always been produced, directly or indirectly, by the burning of fuels. For the past 250 years coal has held pride of place in this respect and annual world production now amounts to some 2 000 million tonnes. In the last half century, however, the contributions of oil, and more recently, natural gas have become of increasing importance.

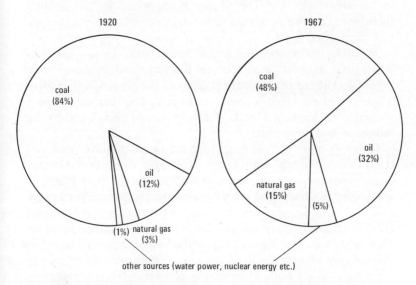

Fig. 2.1 *The changing pattern of world energy sources*

It has to be borne in mind that the known reserves of recoverable coal and lignite, which amount to some $2 \cdot 3 \times 10^{12}$ tonnes or over 900 times the present world consumption, far exceed the combined reserves of all other fossil fuels such as petroleum, natural gas and oil shale. For this reason coal will almost certainly hold its own as a fuel in the foreseeable future.

THE ORIGIN AND NATURE OF FOSSIL FUELS

The conversion of plant material to peat, lignite ('brown coal'), bituminous coal, anthracite and finally the dense, black mineral called jet is usually regarded as being the result of decomposition brought about partially by micro-organisms and partially by high pressures and temperatures produced within the earth's crust. Humic acids, which are soluble in alkalis but insoluble in water, are produced in the presence of air during the enzymatic breakdown of dead plant tissues by soil micro-organisms. The fact that freshly mined coal does not contain humic acids suggests that it is produced in the absence of air. This theory is supported by the gradual increase in the carbon/oxygen ratio which occurs during the transition from wood to jet.

The only environment in which these conditions are found is where luxuriant plant growth occurs in swampland. In these waterlogged conditions there is partial decay of the plant debris to form a humic acid gel which is converted to peat. The alteration of peat to coal is accomplished by the heat and pressure developed by the sediments laid down above it.

Different *types* of coal (e.g. boghead or cannel) are produced from different plant components. The chief distinction between coals is, however, in their degree of coalification (i.e. from lignite to anthracite) in which there is a progressive increase in the percentage of carbon and decrease in the volatile matter released on heating: these differences determine the *rank* of the coal, anthracite being highest in rank, and the majority of the bituminous coals found in this country being medium or low rank. Another major difference, superimposed on rank, is in caking properties. Only a relatively narrow band of low- and medium-volatile coals have the property of forming a hard and strong coke on heating in the absence of air (carbonizing), and these are essential for the manufacture of metallurgical fuel. A slightly wider range of coals are suitable for carbonizing to make gas, and give a softer coke. The remainder of the bituminous coals are not suitable for making coke or gas by

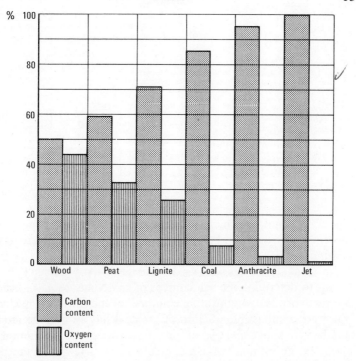

Fig. 2.2 Proportions of oxygen and carbon present in wood and derived fossil forms

conventional methods, and are used for electricity generation, industrial steam raising and domestic fires. Since these high-volatile coals produce smoke on combustion, however, their use as house coals is diminishing and processes have been developed to convert them to smokeless fuels. Anthracite, being very low in volatiles, is naturally smokeless. The *grade* of coals is a marketing rather than a scientific assessment, based mainly on size and ash content.

The mineral content of coal produces incombustible ash which sometimes fuses to form layers of clinker. The loss of heat and cost involved in ash removal is an important economic factor in coal-fired boilers, although in the older types of furnace an ash content of about 7% is desirable to prevent fire-bar damage by overheating.

Rank	Volatile matter %	Proportion of coal mined %	Proportion of heat from volatile matter %
HIGH			
Anthracite	<8	2	5–14
Semi-anthracite	8–14	2	14–21
Low-volatile bituminous	14–22	9	21–28
Medium-volatile bituminous	22–31	7	28–36
↓ High-volatile bituminous	>31	80	36–47
LOW			

The use of moving 'chain' grates and coal injection burners, and more recently fluidized bed techniques, enables ash to be removed continuously.

An X-ray device has been used at the UKAEA, Wantage Laboratories, to determine the ash content of coal. The 'back scatter' of X-rays depends upon the atomic number of the scattering medium. Since the combustible fraction of the coal has an average atomic number of about 6 and that of the ash an average atomic number of about 11, an estimate of ash content can be determined in a few minutes. The presence of iron in the coal is a complication, but compensation for this factor has been achieved by placing a thin aluminium foil X-ray filter over the detector window. The need for continuous monitoring devices for ash and moisture, especially at coke ovens, is increasingly recognized, and several such devices are available, but further research is needed to improve their accuracy and reliability.

Coal has a complex chemical form and contains structures with molecular weights as high as 3 000. These are for the most part clusters of aromatic nuclei fused together into small sheets which are in turn stacked like saucers one above the other. The outer hexagonal rings often possess short side chains which, although usually of the paraffin type, also contain nitrogen, oxygen and sulphur atoms. The layered plate-like structure of the coal molecules is responsible for the hard brittle nature of coal as opposed to the fluid character of petroleum with its long open hydrocarbon chains.

Fuels

(a) Hypothetical structural unit of the type found in crude petroleum

(b) Hypothetical structural unit of the type found in coal

The source of the ring structures found in coal is believed to be the lignin present in the woody material from which the coal originated. The lignin in turn is thought to be a polymer of coniferyl alcohol.

Several theories have been put forward to account for the origin of crude oil. It seems probable that it was formed from the decomposition of marine organisms, both animal and vegetable, which sank to the sea bed in shallow water and became covered by river mud. In the absence of oxygen, decay was inhibited and chemical changes occurred as the result of the pressure and higher temperatures in these source beds, together with bacterial action. As a result the soft body parts of the dead organisms were converted into oil and gas. Pressure of overlying rock layers then forced the oil and gas in all directions along fissures and through porous strata until either it became trapped underground or seeped from the surface. Thus accumulations of oil are usually found trapped in certain geological formations such as anticlines or beneath salt domes, or absorbed by sand or cracked limestone ('reservoir rock'). The

Fuels, Explosives and Dyestuffs

Suggested mechanism for lignin formation from coniferyl alcohol

(coniferyl alcohol → dehydrogenation → dimer of coniferyl alcohol → repeated condensation → lignin)

'perpetual fires' at Baku on the Caspian Sea, which attracted fire worshippers as far back as 600 BC, and the great pitch lake at Trinidad are examples of seepage. The destruction of Sodom and Gomorrah in Biblical times has been attributed to the explosion of petroleum gas released by an earthquake—'the smoke of the country went up as the smoke of a furnace' (Genesis 19: 28).

Crude petroleum usually has an unpleasant smell due to the presence of sulphur compounds and may vary considerably in viscosity, colour and composition.

Paraffin base crudes consist mainly of paraffin hydrocarbons (alkanes) and contain little asphaltic material but appreciable amounts of paraffin wax and lubricating oil fractions.

Asphaltic base crudes contain little paraffin wax but large amounts of asphaltic material and naphthene hydrocarbons (cyclic paraffins).

Mixed base crudes are mixtures containing paraffin wax and asphaltic material together with naphthenes and paraffins.

Only small deposits of of oil have been found in the UK, the first successful drilling being made in December 1919 at D'Arcy near Dalkeith. During World War II 380 wells were sunk, of which 250

Fuels

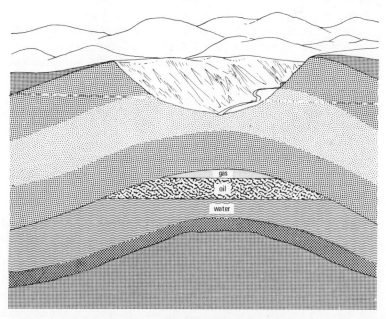

Fig. 2.3 Classic oil trap: an anticline

were producing. During the 1950's further oil strikes were made by British Petroleum at Plungar (Leics.), Egmanton (Notts.) and four sites in the region of Gainsborough (Lincs.). During 1959 oil was discovered at Kimmeridge on the Dorset coast, the well being located on a clifftop overlooking the English Channel. Since 1959 BP have produced 2 million tonnes of oil from English oilfields.

The giant Ekofiskfield in the Norwegian area of the North Sea was discovered in 1970 after exploratory drillings by Philips Petroleum. One of the largest oilfields in the world, it has an estimated area of 55 square kilometres (22 square miles). There are plans to build a pipeline from the field to the Scottish coast. At the moment virtually all the crude oil refined in this country is imported from North America and the Middle East.

Natural gas, which is usually associated with oil deposits, consists mainly of methane together with varying amounts of nitrogen, hydrogen sulphide, ethane, propane and butane. Almost all the natural gas from the North Sea is free from hydrogen sulphide and is of good quality, consisting of 95% methane and 4% nitrogen. The gas from the Dutch Slochteren field is similar but contains 14% nitrogen. The four main North Sea fields have an estimated reserve of 700 hm^3 (25 billion (25 × 10^{12}) ft^3) of gas compared with the 1 680 hm^3 (60 billion ft^3) of Slochteren. Although of high calorific value, this gas is 'dry' and does not contain any appreciable amount of ethane or liquid hydrocarbons as does the 'wet' gas produced in parts of Texas and Louisiana. For this reason the gas is not generally suitable for the production of petroleum chemicals except in the manufacture of synthesis gas. It is estimated however that by 1970 some 9 hm^3 (320 million ft^3) of natural gas could be used daily for ammonia and alcohol synthesis. This will save petroleum feedstocks and enable greater exploitation of Britain's natural resources.

Methane is also produced in coal deposits ('fire-damp') and in masses of decaying vegetation or sewage decomposing in the absence of oxygen ('marsh-gas'). Burning jets of natural gas have been described at least as far back as Roman times—Plutarch (60–140 AD) describes a perpetually burning flare over the entrance to the Egyptian Temple of Jupiter Ammon, and the Chinese transported natural gas using hollowed bamboo pipes. Deposits of petroleum and gas in Egypt were also described in the seventeenth century by the Jesuit priest Athanasius Kircher who commented upon its use by the priests who 'connect an oil deposit by a secret duct with one or more of the lamps provided with wicks of asbestos'.

The first commercial source of natural gas in the UK was discovered near the Midlothian village of Cousland and since 1957 has been fed into the town gas grid at the Musselburgh works. The use of methane drained from coal mines was first attempted on the Continent in 1821 and small plants are in operation to utilize this fuel source at several collieries.

FUEL PROPERTIES OF COMMERCIAL IMPORTANCE

The calorific value of a solid or liquid fuel is the heat liberated when unit mass is burnt. This is determined using an apparatus known as a bomb calorimeter in which the fuel sample is ignited in an

Fuels

atmosphere of oxygen using a hot platinum wire. In Britain the standard heating unit is at present the British thermal unit (Btu). As this unit is rather small, the therm (100 000 Btu) is commonly used. After metrication it is possible that the megajoule (MJ) will be used (1 therm = 105·506 MJ).

Fuel	Average calorific value		Fuel	Average calorific value	
	Btu per lb	MJ/kg		Btu per lb	MJ/kg
Gasoline	20 000	47	Bituminous coal	14 000	33
Diesel oil	19 500	45	Coke	13 000	30
Alcohol (ethanol)	12 000	28	Peat	10 000	23
Anthracite	15 000	35	Dry wood	8 000	19

Calorific values of common solid and liquid fuels

The calorific value of a gaseous fuel is determined by measuring the heat produced by burning a standard volume in air. A specially designed burner is used to heat a constant stream of water passing through a heat exchanger. Town gas is still at present adjusted to the traditional value of 18·63 MJ/m^3 (500 Btu per ft^3) but the calorific value of natural gas is double this value. For this reason natural gas cannot be used with burners designed to use town gas. If this were done, combustion would be incomplete and the flame would be sooty and 'floppy'. Conversion to natural gas involves raising the gas pressure and narrowing the injector so that the volume of gas in the gas/air mixture is reduced by half. Unfortunately the burning velocity of natural gas is low compared with that of the hydrogen-rich town gas and at the high pressure used the flame tends to 'lift off' the burner. This is the reverse of 'striking back' which occurs when the gas pressure is too low. Natural gas burners therefore have a small low pressure 'retention flame' just below the main flame to prevent lift. The flame speed of gaseous fuels is measured in Weaver units—the velocity of the hydrogen flame being taken as the standard value of 100 W.U.

In the case of coal the calorific value is often of secondary importance to the rank, i.e. the proportion of volatile matter. Thus the slow-burning small-flame anthracites are suitable for enclosed central heating boilers which rely on direct heat transfer from the fuel bed. For kilns or steam-raising plant, however, the big flame of the lower rank coals is desirable. Fluidized beds and cyclone burners also require the more readily combustible bituminous coals. Another property of coal which is important in carbonizing processes is its coking characteristics. Mixtures of low- and medium-volatile coals are used to ensure that the charge swells and cokes without becoming wedged in the retort. The specification for metallurgical coke is naturally strict in terms of strength, ash and sulphur content and this limits the coals that can be used to make it. Phosphorus matters comparatively little now that the acid open hearth is disappearing; sulphur in coke is extremely important, because of the extra limestone that must be added in the blast furnace to remove it, which results in loss of heat in the slag and so on. Research in the past two or three years has established that by careful selection blends of different coals can be used, with additives such as coke breeze or anthracite to make good quality coke from coals hitherto considered to be outside the range of coking coals.

An important property of gasoline fuels is their octane rating, a number representing the percentage blend of iso-octane (octane number 100) with n-heptane (octane number 0) which exactly matches the given fuel. Thus a gasoline mixture (i.e. 'petrol') which matches a blend of 90% iso-octane and 10% n-heptane is described as 90 octane. Some aromatics used in internal combustion engine fuels have a performance superior to pure iso-octane. Premium fuels containing these substances may therefore have an octane rating which is above 100. In the UK a star symbol is used to indicate approximately the octane rating of retailed petrol. This ranges from two star (90 octane) to five star (100 octane).

Just as gasoline engine fuels are rated by an octane number, the diesel fuels are determined by a cetane number, which reflects the readiness with which the fuel ignites when injected into the cylinder. Cetane ($C_{16}H_{34}$), which is superior in ignition quality to any of the diesel fuels, is given a rating of 100, while α-methyl naphthalene, which has very poor ignition properties, is rated 0. Blends of these two compounds can be used to standardize diesel fuels.

(1) SOLID FUELS

Although wood, peat and later charcoal were the earliest solid fuels in common usage, opencast and sea coal have probably been used in Europe from Roman times. The combustible 'black rocks' seen by Marco Polo during his travels in the Far East indicate that the Chinese were familiar with coal long before this. Coal was probably first mined in Europe by the Germans and French in the tenth century. Although mining in Britain did not begin for about another 150 years, production had reached 2 million tonnes a year by the time of the Great Plague in the 1660's and quadrupled within the next 100 years. The Industrial Revolution of the eighteenth century produced an enormous demand for coal and by 1860 world production had reached the staggering total of 134 million tonnes. It has been calculated that today in the industrial countries of the world the annual consumption of coal is about 2·5 tonnes per head of population, total world production being over $2\frac{1}{2}$ thousand million tonnes a year.

About three-quarters of the coal mined today is burnt as a fuel for steam raising, heating, smelting, firing kilns and other industrial uses. Most of the remainder is heated to produce gaseous fuels or coke, the tars and oils produced during carbonization providing a rich source of organic chemicals. Small amounts of coal are used as filtration media, ion exchange granules and pigments. A number of processes have also been exploited on a small scale for the production of oil from coal by hydrogenation under pressure. Large scale research is being carried out in the USA at the present time on such processes.

FLUIDIZED COMBUSTION

A recent advance in the utilization of coal as a fuel in large industrial steam-raising plant has been the development of experimental fluidized burners. These are expected to have many advantages over the conventional type of pulverized fuel boilers. Initial and maintenance costs are expected to be lower and poorer grades of coal can be used with reduced air pollution. The fluidized bed is produced by blowing air through an inert layer of hot ash to which crushed coal is added continuously. The temperature of the ash is maintained by combustion of the fuel at about 800°C, the excess ash overflowing from the bed. A unique feature of this type of burner is that the steam-raising tubes are actually buried within the combustion bed. This not only gives excellent heat transfer but cuts down the severe corrosion problems which result from the superheater tubes in traditional furnaces being exposed to gases at temperatures of the order of 1 500° C. In fact one of the major features of fluidized burners is that cheap low-grade coal with a high corrosive quality can be burnt without plant damage. The Central Electricity Generating Board and National Coal Board have developed an ash bed combustor to function at atmospheric pressure. It has been calculated that this would reduce the cost of generating electricity by two-thirds to a figure of £1 per kW. The British Coal Utilization Research Association have also designed a pilot scale pressurized burner to be operated at 1–2 MN/m^2 (10–20 atmospheres) pressure. The combustion chamber is a vertical cylinder about 1·2 m (4 ft) high containing a fluidized bed about 0·6 m (2 ft) in depth. The hot flue gases are blown out through a bank of heat exchanger tubes and then a baffle containing a grit arrester. Conduction within the bed is fast and complete with very little carbon monoxide passing into the stack. The unit uses around 180 kg (400 lb) of coal an hour and produces a heat release of some 9·77 GJ/m^2 (800 000 Btu/ft^2) of bed base.

AIR POLLUTION

The passing of the Clean Air Act in 1956 after the Beaver Report on air pollution has already been mentioned. This legislation was an attempt to remedy what had long been a social evil. The health of town dwellers had long suffered from the pall of smoke which hung over them for much of their lives, obscuring the sunshine and poisoning their lungs. John Evelyn in his *Fumifugium* of 1661 commented: 'London's inhabitants breathe nothing but an impure and thick mist accompanied with a fuliginous and filthy vapour.' Early attempts to remedy the situation failed largely because of the

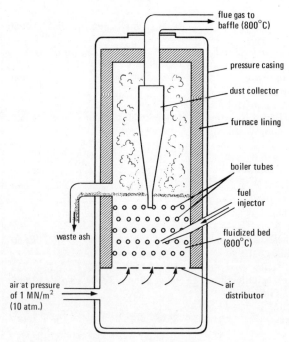

Fig. 2.4 BCURA fluidized bed combustor

lack of suitable smokeless fuels. Coke and anthracite were the only types available in large tonnages, although production of Phurnacite and Coalite had begun before World War II. It was also difficult to persuade consumers to use the more expensive smokeless fuels and finance the conversion of grates where this was necessary. However the disastrous London 'smog' of 1952 which caused the deaths by bronchitis of some 4 000 persons stung the government into action. The gas and coal industries also responded by stepping up the manufacture of smokeless products. This more than offset the shrinking production of gas coke.

The 1956 Act not only gave local authorities powers to create Smoke Control Zones and recompense householders fitting modern grates but enabled them to exercise stringent control over industrial premises. The flues of furnaces consuming more than one tonne of fuel per hour had to be fitted with smoke filters to trap dust particles

	1955	1961	1963	1965	1966	1967
Premium smokeless fuels		(thousands of tonnes)				
(a) Gas industry (including Clearglow, Phimax and Multiheat)	—	522	853	770	800	714
(b) Coal Board and others (including Warmco, Homefire, Coalite and Rexco)	410	644	700	938	1 046	1 181
Total	410	1 166	1 553	1 708	1 846	1 895
Other smokeless fuels		(thousands of tonnes)				
(a) Anthracite and low-volatile coals	1 464	1 571	1 669	1 579	1 850	1 593
(b) Phurnacite	269	655	702	753	733	733
(c) Coke (including hard coke, Gloco and Sebrite)	3 508	2 966	3 591	3 658	3 608	3 529
(d) Others	35	102	124	136	185	180
Total	5 276	5 294	6 086	6 126	6 376	6 035
Grand total	5 686	6 460	7 639	7 834	8 222	7 930

Production of smokeless fuels in Britain (1955–67)

and have their design approved. This legislation has been remarkably successful in achieving its end and fogs and smoke palls no longer plague the cities of Britain. The most serious remaining pollution problem involves the production of sulphur dioxide gas from burning coal and oil. Special smoke washing plant using Thames water has been installed at the Bankside and Battersea power stations in London but the cooling effect on the smoke poses further problems of dispersal from the chimneys. The problem of sulphur dioxide pollution has been eased by the increase in thermal efficiency of power stations and other steam-raising plant. Widespread use of fluidized bed burners should reduce pollution even more.

SMOKELESS FUELS

When coal is heated up to 300° C a certain amount of decomposition occurs but the physical form of the coal remains unchanged. Above this temperature the coal is first deformed and then becomes a plastic mass, while most of the tarry and other volatile matter is driven off. During the final phase up to 1 000° C the remainder of the volatile matter together with hydrogen, oxygen and some of the carbon is expelled and coke is left as a residue. By stopping carbonization at the plastic stage a semi-coke can be produced which is smokeless and easier to ignite and burn in an open grate than coke. The production of these semi-cokes is an important feature of the modern coal carbonization industry. A considerable proportion of the research expenditure of the Coal Research Establishment of the National Coal Board in the last few years has been directed towards the development of new fuels of this type.

A pioneer of the domestic type smokeless fuels was Coalite which was patented in 1890. The first commercial plant was built at Wednesfield in 1907 by the British Coalite Co. Owing to technical difficulties and lack of public interest in the new fuel, however, the company failed. In the twenties production was restarted by the Low Temperature Carbonization Co. and a modern installation was completed in 1927 for large-scale manufacture of smokeless fuel. Low-grade weakly coking coals were carbonized at a temperature of from 450–500°C in narrow, vertical, gas-heated retorts.

The fuel took some time to establish itself and the 'phenolic' tar produced as a by-product was unpopular with the distillers. Nevertheless the easy lighting properties of Coalite, its intense radiant heat output and uniform rate of burning coupled with its smokeless nature, eventually established its use and led to the development of similar fuels.

In 1935 the first Rexco plant was erected in this country by the National Carbonizing Co. on a site adjacent to Mansfield Colliery. The process was based on a method for treating oil shale developed in the USA in the early thirties, and by 1954 production had increased to 30 000 tonnes a year. Most carbonization methods use an outer mantle of burning gas to heat the retort housing the coal charge. The essence of the Rexco process is that the coal charge is heated directly to a temperature of 700–750°C by hot gases drawn through the retort.

The Rexco retorts are large steel vessels about 7·6 m (25 ft) high lined with refractory bricks and holding a charge of 35 tonnes of small weakly coking coal of a type common to the coalfields of Nottingham, Derby, Northumberland and Scotland. Hot gas from adjoining combustion chambers is passed downwards through the coal charge for about seven hours until the top two-thirds is carbonized. The hot gas is then shut off and cold unburnt gas is blown down through the coal charge, carbonizing the lower layer and cooling the whole to a temperature of about 100° C. The complete cycle takes about 17 hours.

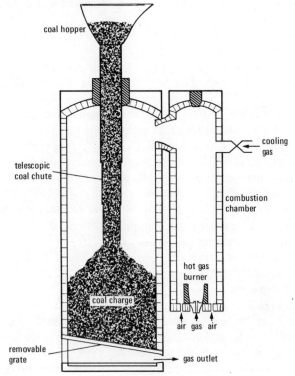

Fig. 2.5 Rexco retort and combustion chamber

A large Rexco plant at Edwinstowe (Notts.) which came into full operation in 1959 has two benches of seven retorts. This installation handles 780 tonnes of coal a day with an output of 535 tonnes of Rexco. A 35 tonne charge of coal yields about 26 tonnes of Rexco, 28 hl (630 gallons) of tar oil and 2·95 GJ (28 therms) of gas. The tar produced by this process is more 'primary' (i.e. less cracked) than low temperature tars formed by external charge heating. Thus naphthalene, anthracene and light oil are virtually absent, but there is a high proportion of phenolics, together with paraffins, olefins and alkylated aromatics. Two rot-proofing agents 'Rexcopine' and 'Rexcote', which have been prepared from the tar oil, have proved very successful especially in the treatment of 'heavy' articles such as coal sacks and ropes.

A typical modern installation for the low temperature carbonization of coal in the traditional externally heated retort is the Eastern Gas Board solid smokeless fuel plant at St Albans. Like the Rexco plant mentioned above, this installation came into operation in the autumn of 1959. The fuel is marketed as Cleanglow, Gloco being also manufactured on the same site. Cleanglow is made in continuous vertical retorts from low-rank Midland coal with an ash content of below 5%. The low swelling and coking properties of this coal enable it to pass rapidly through the retorts and to be extracted by a helical screw with the minimum of breakage and dust production. The carbonizing plant consists of sixty-four retorts in two benches of thirty-two. Each bench is supplied with producer gas by two hot-gas producing units. The Cleanglow is continuously removed from the retorts by a mechanically propelled travelling coke chute. The chute feeds two belt conveyors with a capacity of 35 tonnes an hour and the fuel is then automatically packed into paper bags holding 13 kg (28 lb). In addition 140 000 m^3 (5 million ft^3) of gas per day is fed into the Watford/St Albans grid after electro-detarring and the removal of ammonia in specially designed 'Multifilm' washers.

Solid smokeless fuels can also be prepared from low-volatile coals in the form of small moulded lumps known as 'briquettes'. The first briquette type fuel was manufactured by the Powell Duffryn Company at Bargoed in 1938 by heating anthracite chips bound with pitch to a temperature of around 750°C. The production of this fuel, known as Phurnacite, was taken over by the National Coal

Board in 1942 and the plant at Aberaman now produces some half a million tonnes annually.

Research was then carried out to see if the briquetting process could be used to produce solid smokeless fuel from low-rank coal containing as much as 30–40% volatile material. The problem was to reduce the volatile content of the coal to prevent smoke formation and at the same time retain sufficient of the active constituents to give an easily lighted bright fire with a high radiation factor.

It was found that moderate heating of low-rank fuels not only drives off a proportion of the volatile matter but appears to change the remaining volatile portion of the fuel in such a way that it does not form smoke. Thus a heat-treated fuel of this kind will burn without smoke with a volatile content as high as 23%, whereas a raw coal of 20% volatiles is smoky.

Initially, fuel briquettes made from low-rank coals were bound with pitch, as in the Phurnacite process, and then heated to 200°C. At this temperature the reaction becomes self-sustaining and the heating blast of hot air can be turned off. After several hours the carbonization process is complete, resulting in a high yield solid product of 20–30% volatility.

More recently a smokeless briquetted fuel made from Welsh anthracite has been marketed by the National Coal Board (Multiheat). This is produced at Cardiff by curing pitch-bound briquettes in a bed of sand which is fluidized and kept at a temperature of about 380°C by a hot air blast. The smoky volatile material which is driven off during this stage is used to heat the air blast supplied to the bed. The briquettes are moved along the kiln by increasing the air flow for a 3-second interval. During the remainder of the $1\frac{1}{2}$-minute cycle the bed is kept in a state of 'teeter' (i.e. 'simmering' rather than 'boiling').

Experiments carried out at the National Coal Board Research Establishment at Stoke Orchard have also shown that carbonization takes place very rapidly if the coal is powdered before heating. This reduces the processing time from hours to minutes and can be satisfactorily carried out by blowing very hot air through a bed of coal dust. This process is termed fluid bed carbonization, the heat developed in the reactor being supplied by part of the coal charge burning away. Semi-coke or 'char' produced in this way is plastic when hot and can be extruded (Homefire) or pressed into briquettes (Roomheat) without the use of a binder.

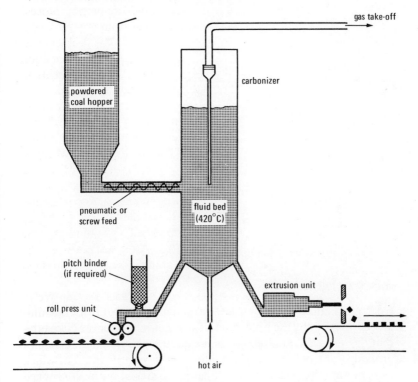

Fig. 2.6 Fluid bed carbonizer showing alternative formation of briquettes

A pilot plant to produce 120 tonnes of briquetted char daily by the fluid bed process was built next to Birch Coppice Colliery in Warwickshire and confirmed that the process was suitable for production on a large scale. The advantages of the fluid bed process are cheapness, ease of control, rapid throughput, and lower capital outlay than for more conventional plant.

A large-scale plant for the manufacture of up to 750 tonnes a day of Homefire briquettes was commissioned in Coventry during 1965. A smaller plant for the production of Roomheat was commissioned during the following year at Markham (Notts.).

	thousands of tonnes				
	1950	1955	1960	1965	1967
(a) Production					
Coal used	1 335	1 651	1 446	949	1 030
Phurnacite	169	331	640	814	827
other briquettes	1 229	1 351	772	93	132
Total	1 398	1 682	1 412	907	959
	1950	1955	1960	1965	1967
(b) Disposal					
Domestic	520	764	813	844	864
Railways	314	571	568	—	—
Commercial, public utilities, etc.	199	95	48	49	66
Export	388	272	80	24	17

Production and disposal of briquetted fuels in Britain

GAS MAKING FROM COAL

Coal gas manufacture by the destructive distillation of coal was first carried out by the Frenchman Tardin in 1618 after the discovery that the 'fire-well' at Grenoble was caused by a smouldering underground seam of coal. Later in the same century the Yorkshire clergyman John Clayton, investigating stories about a ditch containing 'inflammable' water, discovered burning gas issuing from a shallow coal deposit. He describes subsequent experiments on the gas produced by heating coal in the *Transactions* of the Royal Society for 1739. 'At first there came over only Phlegm, afterwards a black Oyle, and then likewise a Spirit arose, which I could noways condense. . . . I observed that the Spirit which issued out caught Fire at the Flame of the Candle and continued burning with Violence as it issued out.'

Although Clayton filled bladders with the gas and lit jets escaping from pinholes in the skin, the credit for using coal gas commercially must go to William Murdoch. In the autumn of 1792 Murdoch illuminated his office in Cross Street, Redruth, using a gas lighting plant of his own design. In 1802 he created a stir by illuminating the Soho Engineering Works in Birmingham with lighted gas jets to celebrate the Peace of Amiens and three years later Pall Mall became the first street in the world to be lighted by coal gas.

Despite initial opposition to the new illuminant from members of the public who were apprehensive of the explosive nature of coal gas and its unpleasant smell, by 1829 there were over 200 gas companies in the UK, the largest being the London-Westminster Chartered Gas Light and Coke Company, founded in 1812 by a German speculator Frederick Winsor.

The early carbonizing processes for the production of coal gas changed very little over the years, the only major alteration being the introduction of the continuous vertical retort—the older horizontal version now being obsolete. The destructive distillation of coal in retorts of this kind is carried out by heating with producer gas or other gaseous fuel at a temperature of about 1 400°C. The charge coal is heated for a period of about twelve hours. Most of the volatile components in the coal are driven off by this time, leaving a residue of coke at about 1 000°C.

Coal is continuously fed into the top of the retort from which the gas is also taken off, and the charge slowly moves down through the retort, which is lined with refractory bricks, to emerge some 7·6 m (25 ft) below as coke, which is steam quenched before extraction. The controlled injection of steam increases the yield of gas by the water gas reaction.

$$\underset{\text{steam}\quad\text{coke}}{H_2O + C} \xrightarrow{\text{heat}} \underbrace{CO\uparrow + H_2\uparrow}_{\text{'water gas'}}$$

Intermittent vertical retorts are also used in which the coal charge of about 13 tonnes is stationary during the carbonization process and is completely discharged at intervals as in the older horizontal retorts.

As the carbonization process is only slightly exothermic, heat equivalent to as much as 50% of the thermal value of the gas produced must be supplied. This is much in excess of that required in oil gasification. In addition, after cooling and removal of tar, coal gas still contains some 3–4% of impurities such as ammonia and hydrogen sulphide. This calls for a more costly purification than in the case of gas made from a petroleum feedstock which contains a negligible proportion of impurities. In fact the production of coal gas by carbonization in retorts is now uneconomic and the process is becoming obsolete. The changing pattern of town gas manufacture is shown by the following table.

Fuel gas sources (millions of therms[a])	1961	1963	1965	1967
Coal gas (gasworks)	1 705	1 638	1 269	976
Coal gas (coke ovens)	511	460	435	394
Oil gas	77	198	692	1 660
Refinery gas	154	258	302	328
Liquefied petroleum gas	40	155	372	409
Natural gas	14	31	313	640
Consumption of fuels for gas making (thousands of tonnes)	1961	1963	1965	1967
Coal	22 200	21 800	17 400	13 400
Light oil (naphtha)	547	921	2 207	4 325
Heavy oil	185	268	249	204
Gas-oil	134	93	74	76

[a] 1 therm = 105·5 MJ.

The purification of coal gas by the indirect recovery method takes place in four phases. Initially the hot gas is sprayed with warm ammoniacal liquor and then passed through a battery of water-cooled tubes (condensation) to remove the tar and part of the aqueous liquid which is collected in tar wells. The remainder of the tar is removed by bubbling the gas through a tank of cold ammoniacal liquor (Livesey tar extractor) or more recently by electrostatic precipitation. In the latter process the gas is passed through a charging unit in which the 'fog' of tar particles and other impurities such as flue dust acquire an electrostatic charge. The gas is then passed through an earthed precipitation column—the charged particles being removed as a result of their attraction to the earthed surface of the precipitator.

During the next stage of purification the gas is thoroughly 'scrubbed' to remove the remaining ammonia by pumping up towers countercurrent to a cascade of dilute ammoniacal liquor. Washers which contain a number of revolving brushes partly immersed in ammoniacal liquor are also used. Both types of scrubbing unit present a large liquid-absorbing surface to the gas. Recovery of ammonia from the ammoniacal liquor is carried out by steam distillation. It is then absorbed in a hot saturated solution of ammonium sulphate containing sulphuric acid—the precipitated ammonium sulphate being removed and sulphuric acid added continuously. In addition to this 'indirect' recovery of ammonia there is the 'direct' or 'Otto' process in which the warm gas after tar removal is passed directly into an acid/ammonium sulphate bath.

Hydrogen sulphide and hydrogen cyanide were originally removed by passing the scrubbed gas through purifiers containing hydrated ferric oxide on wooden grids. The resulting iron sulphide was converted to sulphur within the purifier by allowing the semi-purified exit gas from the box to react with the small amount of air present. When the sulphur content reached about 50%, in some cases the oxide was roasted to yield SO_2 for sulphuric acid manufacture, but usually the spent oxide was simply dumped.

(a) $Fe_2O_3 \cdot H_2O + 3H_2S \longrightarrow Fe_2S_3 \cdot H_2O + 3H_2O$
hydrated iron oxide → hydrated iron sulphide

(b) $2Fe_2S_3 \cdot H_2O + 3O_2 \longrightarrow 2Fe_2O_3 \cdot H_2O + 6S$
air

The ammonium thiocyanate and ferric ferrocyanide, formed at this stage by the reaction of hydrogen cyanide with the iron oxide in the purifier, were not recovered. It is interesting to note that in the early days of the gas industry slaked lime was used to remove the hydrogen sulphide from coal gas. When exhausted, this was dumped and created disposal problems, especially as weathering tended to regenerate the hydrogen sulphide. An immense quantity of spent lime was used in road making and fillings in Belgravia and other parts of London in the latter part of the nineteenth century. The Goswell Road and Westminster gasworks alone dumped some

100 000 tonnes of lime annually at Beckton, sufficient to cover 3·64 hectares (9 acres) of land to a depth of 1·8 m (6 ft).

The removal of sulphur from coal gas is now usually effected by liquid purification, as in the Thylox process in which hydrogen sulphide is removed by bubbling the gas through ammonium thioarsenate solution. The initial cost of such plant is low and the sulphur is disposed of more profitably than the spent oxide.

Finally the gas is washed by passing it countercurrent to a spray of petroleum gas-oil which removes the naphthalene and benzole. These products are then isolated by treating the gas-oil with steam, the stripped gas-oil then being re-used ('re-cycled') after cooling. Apart from its commercial value the removal of naphthalene is essential in order to prevent the blockage of gas mains and valves.

High temperature carbonization of a tonne of coal in a vertical retort yields about 280–370 m^3 (10 000–13 000 ft^3) of purified coal gas, 45 litres (10 gallons) of coal tar, 9–13 kg (20–30 lb) of ammonia and $\frac{3}{4}$ tonne (14 cwt) of coke. The gas consists of about 52% hydrogen, 20% methane, 18% carbon monoxide and 6% nitrogen. The proportion of carbon monoxide and hydrogen is higher than in coke oven gas because of the injection of steam during carbonization in the vertical retort. The calorific value of this gas is about 17·5 MJ/m^3 (475 Btu/ft^3).

Coke ovens are used primarily to produce metallurgical coke for the iron and steel industry. They consist of batteries of horizontal ovens which are usually heated by producer gas made on the site by blowing air over red-hot coke. Small installations of this type are not economical to run and modern coking plants often have over 100 ovens. A typical coke oven installation costing £4 million has been recently ordered by the Northern and Tubes Group of the British Steel Corporation. This includes two batteries of 44 ovens, the first of which was due to be operating by August 1970 and the second by May 1971. Each oven will have a capacity of 32·8 m^3 (1 160 ft^3)—the largest yet constructed in the UK. Working conditions in modern plant of this kind have been greatly improved by automated operations and the use of new silica refractories with better heat transfer characteristics. Another feature of modern coking plant is that it is self-contained and is not only able to

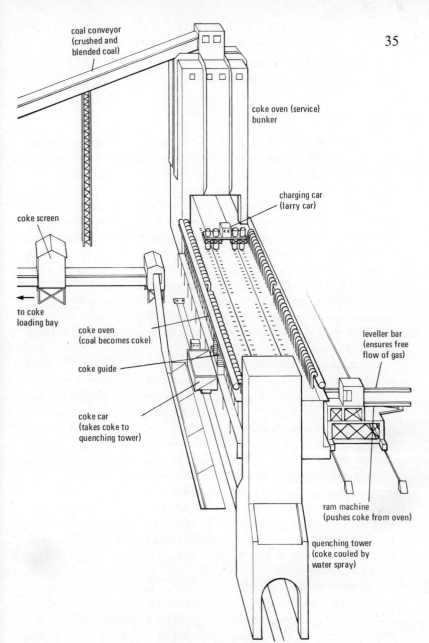

Fig. 2.7 Typical coke oven installation

manufacture producer gas for heating the ovens but can handle its own tar distillation and by-products.

UNDERGROUND GASIFICATION

Between 1949 and 1955 a number of trials were carried out in the UK to investigate the economic possibility of underground gasification of coal. The passage of a current of air through a burning coal seam to produce fuel gas which could be collected at the surface, was suggested in 1868 by Sir William Siemens. Early trials by Ramsay in 1913 were abandoned because of World War I and the first large-scale experiments were carried out by the Russians in 1931. During World War II most of the Russian experimental installations were destroyed, but experiments were restarted after the war. In 1957 a team of British scientists visited several Russian sites, including the large station at Tula near Moscow where gasification of hard coal and brown coal was being carried out. Work on similar lines was meanwhile going on in Italy, Belgium, Poland and French Morocco.

Immediately after the war a report to the Minister of Fuel advised that 'the advantages that might accrue if underground gasification could be successfully and economically carried out are such that experimental work should be started in this country at the earliest possible date'.

As a result, in October 1949 an opencast coal mining site was chosen to start experimental trials at Newman Spinney (near Chesterfield). After preliminary trials and drilling the first useful gas was produced at this site in July 1950.

In order to produce gas from an underground coal seam it is necessary to pass a controlled amount of air or oxygen down to the burning seam by means of a borehole or underground gallery, and then arrange for it to pass through the coal and up to the surface via a second borehole.

The most difficult problem is in the preparation of a channel through the coal to allow free passage for the gas. Until about 1952 this was achieved by driving galleries alongside the coal and then gasifying the coal in these galleries or in borings made in the gallery walls. Later two other techniques were used with success.

The first involved splitting the seam and then burning a channel through it using compressed air or oxygen to join the downcast and upcast boreholes.

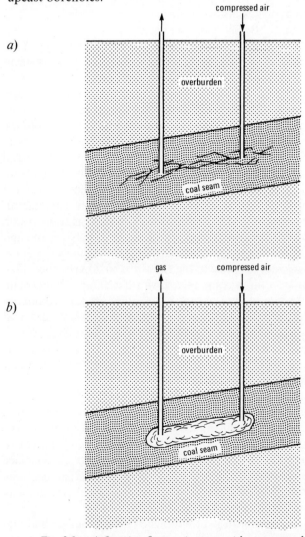

Fig. 2.8 a) Opening fissures in seam with compressed air
b) Passage burnt through seam to connect boreholes

The second technique was to lower graphite or stainless steel electrodes down boreholes into the coal seam at a maximum distance apart of about 30 m (100 ft) and pass a heavy electric current between them. This 'electrolinking' was first tried out in the USA in 1952 and a British working party was invited over to see the process in action during the same year at the Gorgas site in Alabama. Trials in the UK were commenced at the Rockmoor Farm site (Bayton) in 1953 and achieved some success.

Between 1953 and 1955 a modification of the original method was also tried out. This involved drilling for long distances into the coal seam (up to 87 m (290 ft) at Newman Spinney) from galleries or excavated seam faces (directional drilling). The horizontal boreholes thus produced could be used 'blind' or to connect two parallel galleries. Incendiary bombs were initially used to ignite the coal but later Calor gas jets were used.

The gas produced by underground gasification is suitable for direct use on the spot to drive electrical generating plant, or could be piped to town gas grids after processing. It has also been suggested that it could be used as synthesis gas. Unfortunately there is a rapid fall-off in calorific value of the gas as the coal burns away and larger quantities of air become mixed with it making it difficult to maintain a constant quality of output. Another serious problem is leakage of the gas along faults and underground fissures. Because of this, underground gasification in Britain has proved uneconomic and further work has been abandoned.

COMPLETE GASIFICATION

Unlike carbonization, the gasification of a solid fuel involves conversion of the whole of the combustible part of the fuel into gas. In gasification processes of this kind the heat energy required is obtained internally by combustion of part of the fuel, whereas carbonization processes require some form of external heating. Although strongly coking coals which tend to stick together are unsuitable for gasification, one of the attractions of this type of process is in the wide range of solid fuels which can be used in addition to weakly coking coals, such as agricultural waste, peat, lignite, coke and wood.

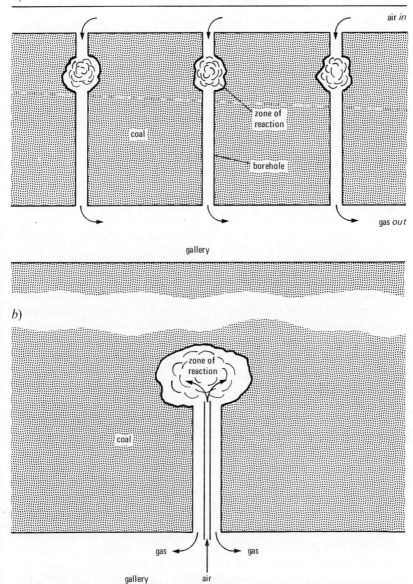

*Fig. 2.9   a) Open boreholes
b) Blind borehole*

Two basic reactions are involved: the strongly exothermic 'Producer Gas' reaction, and the 'Water Gas' reaction which is strongly endothermic and therefore not self-sustaining. By running these two reactions simultaneously or in an alternating cycle, the heat provided by the exothermic process can be made to supply the heat required for the other. In the old-fashioned water gas plant this was usually accomplished by blowing steam and air alternately through a bed of red-hot coke.

Producer gas reactions—strongly exothermic

$$\left. \begin{array}{l} C + O_2 \longrightarrow CO_2 \\ C + CO_2 \rightleftharpoons 2CO \\ 2CO + O_2 \longrightarrow 2CO_2 \end{array} \right\} + \text{heat energy}$$

Water gas reaction—strongly endothermic

$$C + H_2O + \text{heat energy} \rightleftharpoons CO + H_2$$

Since air is used as the oxygen source, producer gas has a low calorific value ('lean' gas) because of its high nitrogen content. This makes it uneconomical to distribute. The problem of nitrogen dilution is avoided in the case of water gas but the gas is still lean and also highly toxic because of its carbon monoxide content. Enrichment of water gas was achieved by using the hot 'air blast' gas to crack naphtha (light petroleum distillate), the resulting oil gas being added to the main product ('carburetting'). Even before World War II, however, experimental work was being carried out to discover a cheap and efficient technique to produce fuel gas with a low nitrogen content by gasification. The 'double gas' plants, which are now obsolete, were an early attempt to gasify weakly coking coal in this way. In effect they consisted of a combined water gas and carbonizing plant, steam being blown into the base of the retort—the resulting water gas and unchanged steam passing up through the carbonizing charge and mixing with the rich coal gas. An experimental G.I. ('Gaz Integrale') plant of this kind was erected at Kensal Green Gas Works in the late 1940's producing gas with a calorific value of 12·3 MJ/m^3 (335 Btu/ft^3).

All the carbonizing and gasifying processes mentioned up to this point are termed 'conventional' and are carried out at low pressure using a solid fuel which is either heated or treated with a hot-air or steam blast. The conventional processes, however, are rapidly

Fuels

being replaced by the so-called 'unconventional' processes which operate at high pressures and use oxygen, oxygen-enriched air, or hydrogen to gasify solid fuel completely. The resulting ash is often liquefied and removed as slag. Already these processes in turn are being replaced by oil gasification and natural gas.

In 1921 two gas engineers at Leeds University, Hodsman and Cobb, proposed the gasification of coal by an oxygen-steam blast. The idea was favourably received by the Germans who were being forced to exploit their deposits of brown coal after World War I because of the loss of the Saar coalfields and lack of funds to buy foreign coal. Early German experiments in 1927 using lignite briquettes were moderately successful and in 1936 the Lurgi Company erected a plant with two generators at Zittau which produced 19 800 m^3 (700 000 ft^3) of gas daily from lignite fuel. During World War II in 1940 a ten-generator installation was built at Bohlen near Leipzig which was soon producing over 71 000 m^3 (2·5 million ft^3) of town gas daily for the Leipzig-Magdeburg grid, and another plant of half this output was erected at Brux in Czechoslovakia. Following the war a Lurgi plant of 1·98 million m^3 (70 million ft^3) was erected at Dorsten (1955).

The success of the Lurgi process attracted the attention of the British gas industry, and experimental work was carried out which showed that at a temperature of 900°C steam under pressure would form methane from bituminous coal, according to the reaction:

$$(a)\ \underset{\text{steam}}{H_2O} + \underset{\text{coke}}{C} \xrightarrow{+\text{ heat}} CO\uparrow + H_2\uparrow$$

$$(b)\ 2H_2 + C \xrightleftharpoons[]{900°C\ +\ \text{high pressure}} \underset{\text{methane}}{CH_4}$$

This was important because of the high calorific value of methane (37 MJ/m^3 (992 Btu/ft^3)). In 1952 the Gas Council and National Coal Board set up a Committee to study the possibility of erecting a Lurgi plant at Westfield near Loch Leven in Fife. After some delay the Westfield plant was commissioned by the Scottish Gas Board and came into full production in 1962 with a capacity of 850 000 m^3 (30 million ft^3) of gas a day. Also in 1952 the West Midlands

Fuels, Explosives and Dyestuffs

Gas Board after field trials decided to erect a Lurgi plant at Coleshill which was finally commissioned in 1963.

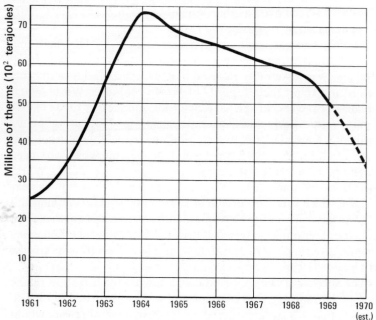

Fig. 2.10 Production of Lurgi gas in the UK

In these Lurgi plants a fixed bed of coal about 3 m (10 ft) deep is gasified by blowing through a mixture of oxygen and superheated steam at a pressure of about 2–3 MN/m^2 (20–30 atmospheres). Oxygen is provided by air-liquefaction plants each capable of producing 100 tonnes of oxygen per day. The coal charge is fed into the top of the gasifier by means of a pressurized hopper, the powdered ash being extracted at the base by a similar device. Lurgi gas contains principally the carbon oxides, hydrogen and a little methane. The crude gas is cooled in a 'waste-heat' boiler which is used for steam production, and benzole removed by oil washing. The debenzolized gas is next passed with steam over a cobalt/molybdenum catalyst which converts the toxic carbon monoxide

Fuels

to carbon dioxide with the simultaneous production of hydrogen (water gas shift reaction).

$$H_2O + CO \xrightarrow{Co/Mo\ catalyst} H_2 + CO_2$$

The carbon dioxide and hydrogen sulphide content of the gas is then substantially reduced by absorption at high pressure in a 35% solution of potassium carbonate at 110°C. This is carried out in a 'Benfield' plant (named after its two American inventors, Field and Benson).

Finally the gas is dried and enriched with natural gas, propane, or butane. As Lurgi gas is produced under pressure it is very suitable for high pressure distribution. Recent developments have made the Lurgi process uneconomic to run in Britain and the existing plants are being phased out.

An interesting application of the Lurgi process is used in the South African 'Sasol' installation at Sasolburg in the Orange Free State. Low grade coal which is very cheap (about 30p a tonne) and plentiful in this area is gasified in Lurgi convertors and after purification is reformed to give synthesis gas (carbon monoxide and hydrogen). This is then converted into petroleum, waxes, alcohols and other organic chemicals by the Fischer Tropsch technique which makes use of iron or cobalt catalysts under a variety of different operating temperatures and pressures. The plant was designed

Aerial view of the Lurgi high pressure coal gasification plant and ancillary installations at the Coleshill works of the West Midlands Gas Board

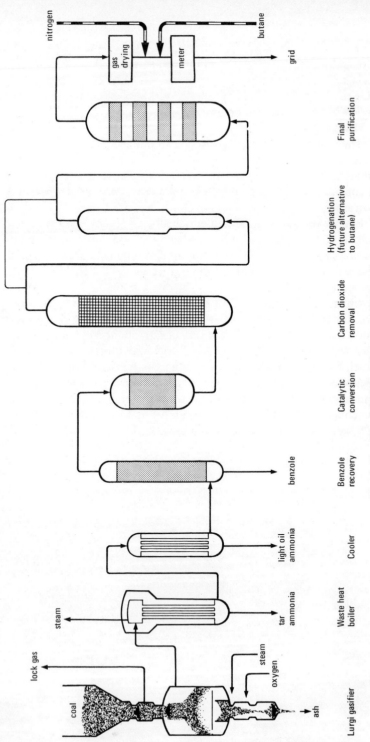

Fig. 2.11 Simplified flow diagram of Lurgi plant

to achieve an output of 250 000 tonnes of synthetics a year but this has not been realized. The possibility of using this process in the UK was considered by a Committee on Coal Derivatives (1960) but rejected as uneconomic.

In addition to the Lurgi process other solid fuel gasification techniques are in use in which the finely divided fuel is carried in suspension in a gas stream or churned up ('fluidized') into a turbulent mass by a gas blast. The Winkler process which was first proposed in 1925 is of the latter type. The finely ground fuel rests on a slotted grate to a depth of $1-1\frac{1}{2}$ m (3–5 ft) and an oxygen-steam blast containing up to 50% oxygen is blown through it. The gas blast creates a number of temporary channels in the powdered fuel and this gives it the appearance of boiling—hence the term 'boiling bed' which is often used to describe this kind of fuel bed. A certain amount of the powdered fuel is carried over with the gas and is oxidized by a secondary oxygen blast. The process can be either continuous or cyclic and is used extensively in Germany and Czechoslovakia.

Other processes have been designed to gasify the pulverized solid fuel by suspending it in an atmosphere of oxygen or air. In the Koppers-Totzec plant, dried powdered coal is blown by a controlled amount of oxygen from steam-cooled nozzles into a combustion chamber. The high temperature of gasification in this type of plant (1 100–1 300°C) produces a gas which is free from tar and liquid hydrocarbons, making it very suitable as synthesis gas.

The Ruhrgas vortex process is similar in principle but relies upon the gasification of coal powder using superheated air at 700°C instead of oxygen. High temperature gasification plants of this type in which the ash is removed in the molten state are called 'slagging producers' and have particularly interested German engineers. They have not, however, attracted much attention in the UK.

A revolutionary method for gasifying coal on an experimental scale was reported in 1967 from the US Bureau of Mines. Research workers used a 2–10 kilojoule source laser beam generating temperatures of over 1 000°C to disintegrate coal samples. The gaseous product produced contained up to 25% acetylene together with lesser amounts of diacetylene, vinyl acetylene, hydrogen cyanide, ethylene and ethane. These products were obtained in

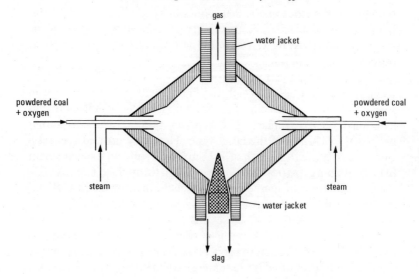

Fig. 2.12 Koppers/Totzec combustion chamber

milliseconds compared with the minutes or even hours required by more traditional methods. This process is more likely to be used for analysis than commercial gas making, however.

PRODUCTION OF OIL FROM COAL

Methods of producing oil from coal have been attempted using direct hydrogenation in order to raise the hydrogen/carbon ratio. In the German Bergius process pulverized coal is mixed with a heavy tar oil and a tin salt catalyst and heated for some hours with hydrogen under pressure. After maintaining the contents of the reactor at 600–800°C for the requisite period of time a thick tar-like oil is produced. This is distilled in the same manner as crude natural oil, fractions being taken off as required. High octane aviation fuel can be produced from coal and other fuels by this method. During World War II some 22·750 hl (500 000 gallons) of aviation fuel a year were produced at an ICI hydrogenation plant at Billingham and a joint ICI/Shell plant at Heysham which used feedstocks of creosote and gas-oil respectively. The process was uneconomic, however, and has been dropped. To supply the present UK total oil consumption of 70 million tonnes a year would require something like 250 million tonnes of coal. Research into the production of oil and gas from coal is still being carried out in the USA with massive government support. This is not only because of pressure from coal producers but also because American oil and gas reserves are thought to be running out rapidly and there is strong opposition to importing oil.

Experience gained in coal hydrogenation has recently been utilized in a novel technique involving the solvent extraction of coal. After removal of ash from the liquid coal extract by filtration, and the extraction of certain volatiles, the residue can be reconstituted to produce an ashless smokeless fuel or a range of plastic pitch-like materials. The extraction need not be complete and can be carried out using a variety of coal-based oils and tars. Thus coal digested in pitch can be used as a road making material.

(2) LIQUID FUELS

PETROLEUM

Until about a century ago, man's acquaintance with petroleum (L. *petra*—rock, *oleum*—oil) was limited to natural seepages. These evaporated to give deposits of bituminous material as in the case of the Pitch Lake at Trinidad, or ignited to produce 'eternal fires' such as those which attracted the fire-worshippers at Baku as long ago as 600 BC. The first well sunk for the express purpose of obtaining crude oil was dug by a Canadian, J. H. Williams, in 1858. On returning from the Baku oilfields he struck oil at 20 m (65 ft) in Black Creek, 160 km (100 miles) or so south of Toronto in what is now known as the 'Chemical Valley' of Canada.

The petroleum industry, however, is usually considered to have begun its eventful history in 1859 when the first mechanically drilled well was sunk by 'Colonel' E. L. Drake, an employee of a lamp oil firm, using seepage crude oil as a raw material. Drake, using a cable tool, came across oil at a depth of 21 m (69 ft) near Titusville in Pennsylvania. The cable tool was simply a length of heavy steel tubing fitted at the lower end with a toughened bit and suspended from a wire rope. Drilling was accomplished by lifting the tool and then letting it drop back down the well shaft, thus driving it deeper and deeper into the ground, the earth and rock being removed at intervals with a hollow tube called a 'bailer'. This technique had been used by the Chinese in drilling for salt as early as the Shu Han Dynasty in the third century AD.

The success of Drake's well started an oil boom which was intensified at the turn of the century by the introduction of the idea of rotary drilling which enabled wells to be drilled more quickly and to greater depths.

The USA is still the largest producer of crude oil and natural gasoline in the world although its contribution has fallen from 72% of the total world production in 1927 to about 33% in 1968. The breathtaking rate of expansion of the petroleum industry in the first century of its existence is indicated by a comparison of the production figures in the USA for 1860 (40 tonnes), 1900 ($1\frac{1}{4}$ million tonnes) and 1950 (40 million tonnes). The consumption of petroleum products in the UK for 1968 is given below. It is interesting to note that the daily throughput of a large refinery such as BP (Kent) Isle of Grain is over 4 000 tonnes of crude—100 times the world production for 1860.

	tonnes
Gasoline (motor spirit)	12 800 000
Kerosine	2 036 000
Diesel fuels (motor vehicles)	4 578 000
Gas (diesel oil)	9 044 000
Lubricating oil	1 134 000
Bitumen	1 830 000
Paraffin wax	55 000

Consumption of petroleum products in the UK (1968)

Crude petroleum is a complex mixture of hydrocarbons, the content of which varies considerably according to its place of origin.

Fuels

Thus crude from Venezuela and Baku is rich in cyclo-paraffins (naphthenes), while Pennsylvanian crude contains some 70% of open chain paraffins and crude from Borneo contains a high proportion of aromatics such as toluene and benzene. In addition many of the American crudes contain large volumes of dissolved natural gas and leave only small amounts of solid residues after distillation. Some of the Mexican crudes on the other hand contain considerable amounts of bitumen, and the proportion of solid residues present in the oil from the Schoonebeek field in Holland is so high that the crude becomes semi-solid on cooling. For convenience the crudes are generally classified as 'Paraffin-based', 'Asphaltic-based' and 'Mixed-base'.

OIL REFINING

The function of an oil refinery is to produce, as economically as possible, the quality and quantity of oil products currently in demand, using the crude oils supplied to it. Thus the refinery has two main tasks: the separation of the constituents of the crude oil by distillation, followed by conversion of these fractions into the products required. It is then necessary to carry out certain finishing processes in order to remove undesirable impurities and provide the customer with an acceptable end-product.

Separation Processes

(a) Distillation

Crude petroleum arrives at the well-head under pressure as a dark brown or black liquid with an unpleasant smell. It varies in viscosity from a thin mobile liquid to a heavy tar-like consistency and usually contains large quantities of dissolved petroleum gas, and smaller amounts of water. Removal of the gas and water is carried out by passing the crude through a flowline to the gathering station. Here the gas is removed in separators and the water drained off in tanks. Occasionally the water forms an emulsion with the oil, and separation can only be achieved by breaking the emulsion chemically by heating or by subjecting the oil to a high alternating voltage of 15 kV.

At this stage the crude is unmarketable and must be initially separated by fractional distillation into groups of compounds within specific boiling point ranges ('fractions' or 'cuts') and a residue of higher boiling point components. A fraction such as gasoline may be required for a specific purpose, other fractions may undergo further separation by secondary distillation or solvent extraction. Alternatively fractions may be blended together to produce a particular boiling point mixture.

The constituents of crude petroleum have been likened to a pack of playing cards, primary distillation sorting out the cards into suits and secondary distillation arranging the suits in numerical order. A process analogous to blending occurs when cards are selected during play to produce a given combination.

Until the turn of the century distillation was carried out by a batch process, the crude being heated in tanks. The kerosine fraction was in great demand at this time, but little use could be found for the highly inflammable and volatile gasoline fraction which was often allowed to run to waste. With the advent of the internal combustion engine more efficient fractionation was required and the continuous still was developed, for use at either atmospheric or higher pressures, or under partial vacuum.

In a modern primary distillation unit, the crude oil is first preheated by passing through heat exchangers. The warmed crude then passes into a specially designed furnace called a pipe still. This is a brick-built chamber which is heated by oil burners set in the wall at one end. The oil is circulated continuously through a long tube formed into parallel loops within the still and is heated initially by the flue gases and finally by the direct radiant heat from the burners. The final temperature of the oil is 300–350°C—higher temperatures are avoided to prevent thermal decomposition taking place.

The hot oil and vapour from the pipe still then enter the primary fractionating column at a point about a quarter of the way up. The column is a tall steel cylinder which is divided horizontally by a number of specially designed perforated metal plates (trays). The perforations in the trays lead into short lengths of tubing (risers) so that condensed vapour collects in a pool upon each tray. The level of the condensate is kept constant by means of an overflow pipe which leads to the tray beneath. Each of the risers is surmounted by a cap (bubble cap) the edges of which are just submerged below the level of the condensate pool.

In this way the hot vapour rising through the perforations in the trays is forced to bubble through the condensate before passing on up the column. This has the double advantage of revaporizing any lower boiling materials present in the condensate and stripping the vapour of its less volatile components by condensation. As each tray is slightly cooler than the one below it, this repeated process of condensation and revaporizing effectively separates the crude feedstock into a number of fractions with different boiling point ranges which can be withdrawn continuously from different levels in the column.

It will be noted that unlike the columns used for fractionating liquid mixtures on a laboratory scale, the large-scale continuous units in commercial use maintain a steady temperature gradient from top to bottom. The temperature gradient can be adjusted by returning part of the condensed fraction to the column as 'reflux' and by controlling the temperature of the heated feedstock. In

Crude oil fractionation tower and ancillary plant in the Shell Odessa refinery (Texas, USA)

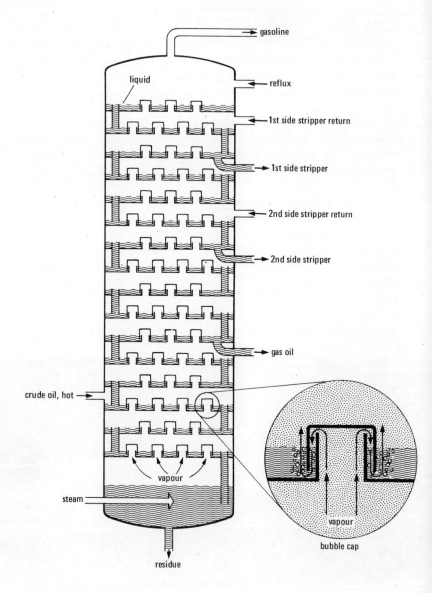

Fig. 2.13   Fractionating column showing trays and bubble caps

addition the high boiling point residue in the base of the still ('bottoms') is continually reheated by circulation through a small auxiliary heater ('reboiler').

With the exception of the light gasoline distillates, the fractions are 'stripped' after removal from the primary fractionating unit. This is carried out in small bubble cap fractionating units in which unwanted volatile components can be vaporized and returned to the main column. Heat is provided by injecting steam into the base of the stripper and by circulating the bottom liquid through steam reboilers. The gasoline fraction which is not returned to the column as reflux is transferred to a light distillate column for further fractionating, the butane and propane being removed under pressure in debutanizer and depropanizer columns.

The fraction from the crude oil distillation plant may be divided into four broad groups, refinery or 'tail' gas, light distillates, middle distillates and residue.

Refinery gas consists mainly of the volatile paraffins methane, ethane, propane and butane. The methane and ethane are used as fuel gases on the refinery site or are piped to local Gas Board reforming units where they can be used for making town gas. The propane and butane on the other hand are liquefied and marketed in steel bottles ('Bottogas') as liquefied petroleum gas (LPG). The LPG is stored on the refinery site in round 'Hortonspheres' or horizontal cylinders.

Light distillates include a low boiling gasoline fraction (light naphtha) (30–120°C), naphtha (120–200°C) and kerosine (250–300°C). After refining, a proportion of the gasoline fraction is directly incorporated into motor spirit ('straight run' spirit), promoting easier starting. The major part of a motor spirit blend, however, is provided from the products of the reforming and cracking processes described later.

The more volatile gasolines (30–80°C) are called petroleum ethers and these are commonly used as solvents for fats and essential oils, and as cigarette lighter fuel. A number of special boiling point (SBP) solvents are also prepared from light distillate components boiling between 35–160°C. These are chosen for special industrial applications, such as in the preparation of rubber solutions or the extraction of seed oils, and are either designated by numbers (e.g. SBP 2) or more usually by their boiling range (SBP 70–95°C).

Solvents known as 'white spirits' with a boiling range of 150–200°C are also produced from the naphtha fraction and contain from 15% to 80% of aromatics. Because of their solvent power and volatility they can be used as turpentine substitutes in thinning paints and cleansing paint brushes, and also for dry cleaning and degreasing raw wool.

Naphtha is the main feedstock for petroleum chemicals in the UK and for cracking processes designed to produce town gas (e.g. ICI steam naphtha process). Kerosine (paraffin oil) is also used as a feedstock for reforming processes, for illuminating and heating, and as a solvent for bitumen. The more volatile part of the kerosine fraction is used as a fuel for tractors and jet aircraft.

Middle distillates provide diesel fuel and gas-oil and they are also blended with certain residual products to produce furnace fuels. Gas-oil is used in the gas making industry for carburetting water gas, for the recovery of benzole and more recently for protein synthesis. Middle distillates also provide the feedstocks for a variety of catalytic and thermal cracking processes.

Residue—this contains substances which are not sufficiently volatile to be removed by distillation at atmospheric pressure without

An installation at Lisbon (Portugal) for the storage of liquefied petroleum gases. The round Hortonspheres hold 600 m³ of propane while the smaller horizontal cylinders have a capacity of 113 m³ and are intended for either propane or butane

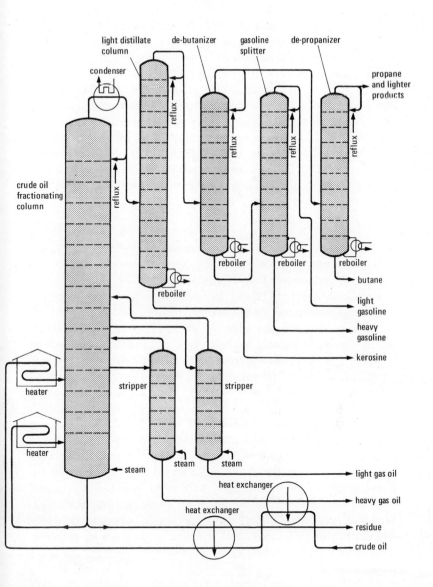

Fig. 2.14 Modern crude oil distillation plant

undergoing thermal decomposition. Distillation at low pressure in a vacuum distillation unit, however, produces a further separation into gas-oil, lubricating oil, a heavy oil feedstock suitable for catalytic cracking and a residue containing bitumen which may also be suitable for thermal cracking or as a fuel.

Specially purified vacuum distillates known as white oils are used in the preparation of cosmetic emulsions, in medicine (liquid paraffin), as coolants and insulating fluids in electrical gear and as special lubricants in cases where ordinary lubricating oils would be undesirable (food processing and textile manufacturing machinery). Purification of the lubricating oil fraction also yields paraffin wax and a mixture of solid and liquid hydrocarbons termed 'Petrolatum' or 'Petroleum Jelly'.

(b) Solvent extraction

Although distillation is the main instrument used for the separation of crude oil into fractions, increasing use is being made of solvent extraction techniques. One of the advantages of this method of separation is that it can be used to extract specific components from a mixture. Distillation, on the other hand, relies solely upon differences in boiling point to effect separation into different fractions which, therefore, contain a whole range of chemical types such as aromatics, naphthenes and paraffins. Solvent extraction was pioneered in 1907 by a Rumanian, Edeleanu, who developed a process for removing aromatic hydrocarbons from kerosine using liquid sulphur dioxide as a solvent.

The Edeleanu process is still used today for the treatment of kerosine. The heavy liquid sulphur dioxide passes down a tower countercurrent to the relatively light kerosine. The purified kerosine or raffinate (Fr. *raffiner*—to refine) is removed from the top of the tower while the sulphur dioxide is recovered from the bottom and recycled after removal of the dissolved aromatic content which is used as a tractor fuel and as a raw material for chemical synthesis. The purification of kerosine in this way is necessary to avoid the undesirable smoky flame which characterizes burning aromatic compounds.

In 1926 the Edeleanu process was applied to the purification of lubricating oils, although furfural and occasionally phenol are now used as solvents in place of liquid sulphur dioxide. Much thought has gone into the design of extraction towers to prevent loss of the costly solvent and to ensure its thorough mixing with the oil.

Another example of solvent extraction is to be found in the removal of asphalt (bitumen) from the high quality lubricating oils ('bright stocks') produced from residual heavy oils which cannot be distilled. Liquid propane under pressure is passed countercurrent to a descending stream of oil passing down the extractor tower. Lubricating oil is preferentially dissolved in the propane and taken off the top of the tower while the bitumen is withdrawn at the bottom.

Liquid sulphur dioxide or diethylene glycol have also been utilized to extract the aromatics from gasoline distillates, the extract being used as a high grade aviation or motor fuel. Similarly reformed gasolines can be separated into high octane (aromatic) and low octane (paraffinic) fractions using solvent extraction methods.

The 'Duo-Sol' process for the separation of aromatic and paraffinic components uses a two solvent system in which one solvent known as 'Selecto' (a mixture of phenol and cresylic acids) is mixed with the oil feedstock, and passed countercurrent to the second solvent which is liquid propane. The raffinate is carried in one direction by the propane whilst the undesirable aromatics are removed by the Selecto. Because of the difficulties inherent in recovering two solvents at once, however, this process is not economical. It is usual to carry out the propane extraction first followed by a second extraction using furfural.

Conversion Processes

The task of the refinery is not merely to separate and blend the components of crude oil. Some of the products may be surplus to demand, others may be present in the crude in insufficient quantities to satisfy the demand. In general the lighter distillates and refinery gases are in greater demand than the heavier, less volatile products. The skill of the refiner lies in matching his product output to demand, irrespective of the nature of the crude feedstocks he is using. This involves conversion of the heavier distillates into the more desirable light fractions, the structural reforming of products to improve their quality, and in certain instances the building up of more complex molecular structures from olefin gases.

(a) Cracking

Cracking describes a process in which larger molecules are broken down into smaller ones either by subjecting them to a high temperature and pressure (thermal cracking) or by using a suitable catalyst (catalytic cracking).

Thermal cracking. If a long chain paraffin hydrocarbon is heated sufficiently strongly the absorbed energy can produce molecular vibrations severe enough to rupture the chain. Reduction in chain length of the hydrocarbon molecules is not the only effect of thermal cracking. As the amount of hydrogen in the parent molecule is insufficient to provide a full complement of hydrogen atoms for the smaller daughter molecules, so the hydrocarbon fragments become more and more unsaturated as cracking proceeds.

$$CH_3(CH_2)_6CH_3 \xrightarrow[2 \cdot 5 \text{ MN/m}^2 \text{ (25 atm.)}]{500°C} CH_3CH=CH_2 + CH_3(CH_2)_3CH_3$$
$$\text{n-octane} \qquad\qquad\qquad\qquad \text{propylene} \qquad \text{n-pentane}$$

The conditions under which thermal cracking is carried out are not usually severe enough to cause the breaking open of ring structures present in the feedstock, although side chains attached to the rings are cracked. Secondary reactions also occur as cracking proceeds, such as the formation of branched chain hydrocarbons. This improves the anti-knock characteristics of the cracked product, an important consideration if it is intended as a motor fuel.

Although thermal cracking is mainly used for the production of gasoline from heavy oil feedstocks, the process is also used for reducing the viscosity of residual fuel oils ('visbreaking') and for the production of detergent base ('Teepol') from paraffin wax.

Most thermal crackers are based upon the Dubbs plant which was originally installed in 1921 at Wood River Refinery in the USA. This type of plant can either be operated to produce fuel oil (non-residual process) or, using higher temperatures and pressure, gasoline with a residue of coke (residual process).

The heavy residual oil is preheated by pumping it into the fractionating column being used to condense the hot cracked vapours.

Fuels

The hot feedstock is then passed into an oil-fired heater to raise its temperature further to 465–490°C when partial cracking occurs. Cracking is completed at a pressure of about 2·5 MN/m² (25 atmospheres) in a cylindrical tower termed the reaction chamber. The cracked oil is then passed through a pressure reducing valve and enters a flash chamber where the pressure is allowed to fall to about 1 MN/m² (10 atmospheres). A high proportion of the hot cracked liquid vaporizes ('flashes off'), any liquid residue being taken off and used as fuel oil. The gaseous cracked products are fed into the base of the fractionating column where they are cooled by the incoming 'feedstock'. The more volatile gasoline fraction is taken from the top of the column and passes through a stabilizer where gases and short chain hydrocarbons are removed. The heavier residual product mixes with the feedstock and is recycled.

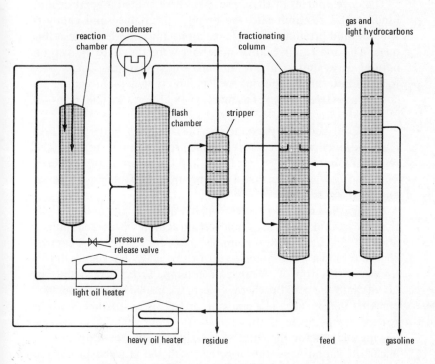

Fig. 2.15   Dubbs process for selective operation

Catalytic cracking. The high temperatures and pressures involved in thermal cracking may be avoided by the use of a suitable catalyst. In addition to the resulting saving in production costs, the gasoline produced by catalytic cracking has superior anti-knock properties due to the formation of aromatic and branched chain structures. There is also a much smaller production of olefins than is associated with thermal cracking techniques, resulting in a more stable product which is unlikely to form undesirable gums.

The first commercially successful catalytic cracking process was developed in 1936, using a fixed catalyst bed. This was followed by the 'Thermofor' or moving bed process in which the catalyst was continuously regenerated and recycled through the plant, and finally by the fluid bed or 'Orthoflow' process in which the catalyst is fluidized by a blast of oil vapour and steam.

The development of these processes necessitated a great deal of fundamental research into the design of suitable solid catalysts which would promote desirable reactions and suppress undesirable ones. The most satisfactory materials were found to be certain natural clays and synthetic alumina/silica mixtures. These were formed into highly porous pellets and powders providing an enormous active surface of as much as 18 hectares/kg (100 acres/lb) of catalyst.

The fixed bed catalytic cracker consisted of a battery of three reactors which were successively cycled for cracking, stripping and regeneration, the catalyst remaining in position for all three phases. The fixed bed technique has a number of disadvantages, however, and has been replaced by continuous processes.

In the 'Thermofor' process which was first operated in 1943, the catalyst is in the form of small cream coloured pellets containing about 12·5% by weight of alumina to 87·5% of silica. The catalyst charge is blown by compressed air or carried by bucket chain to a hopper at the top of the reactor column. From here it falls by gravity into the reactor where it meets the heated oil feedstock and cracking takes place. The cracking process is exothermic, and the reactor temperature is maintained at about 400–500°C. Before leaving the reactor the catalyst is stripped with steam to remove adsorbed vapour and this together with the cracked product is taken off at a point near the base of the bed.

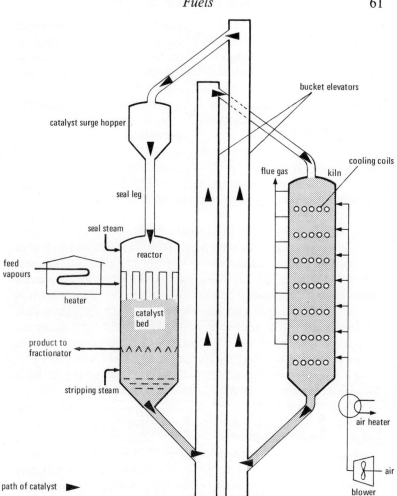

Fig. 2.16 Thermofor catalytic cracking process

During the cracking process the catalyst pellets become coated with carbon and this is removed in a kiln or regenerator. Here hot air is blown into a slowly descending column of catalyst at a dozen or so levels in order to burn off the carbon deposit. The temperature has to be carefully controlled at this stage to avoid damage to the catalyst by overheating, an upper limit of 590°C being maintained. After regeneration the catalyst is lifted to the reactor hopper for recycling. In large 'Thermofor' installations the circulation of catalyst may be as high as a thousand tonnes an hour, during which time some 13 tonnes of carbon have to be burnt off in the regenerating kilns, and a throughput of some 120 tonnes of oil is achieved.

The high catalyst/feedstock ratio limits the efficiency of the 'Thermofor' type plant and it has now been superseded by the 'Orthoflow' type of fluid bed cracker in which there is more effective transport of the catalyst. The first 'Orthoflow' cracker was erected in 1957 at the Tidewater Refinery near Wilmington (USA). Since that time it has been widely adopted and is the only type now used in most refineries.

The catalyst, which is in the form of very fine powder, is circulated through the plant as if it were a fluid. After regeneration by blowing with air to burn off the carbon layer, the hot catalyst is mixed with the oil feedstock which rapidly vaporizes and is carried by a steam blast up into the base of the reactor vessel through a steel tube ('riser'). The cracking process begins while the catalyst/vapour mixture is still in the riser and is completed in the fluidized bed of catalyst within the reactor. The cracked vapour is taken off the top of the reactor into a fractionator while the spent catalyst is continuously withdrawn and is blown up a second riser to the regenerator by a steam blast as before. Steam is also used to strip the catalyst of traces of oil before it passes by gravity feed to the regenerator, and helps to fluidize it.

To prevent catalyst being blown out of the top of the reactor or regenerator with the cracked vapour and flue gases, dust traps (cyclones) are fitted to remove dust particles from the vapour streams. Nevertheless a certain loss of catalyst occurs and a 'topping up' reserve of about a third of the plant capacity is usually held. The working load of catalyst varies from one installation to another, but charges of 2 000 tonnes or more are used in the largest units.

The Shell Petroleum Co. have recently developed a modified fluid bed cracking process in which the cracking is carried out in two separate stages. The oil feedstock is first passed through a preliminary cracking stage in which a certain amount of refinery gas and gasoline is produced. The unreacted material is then passed into

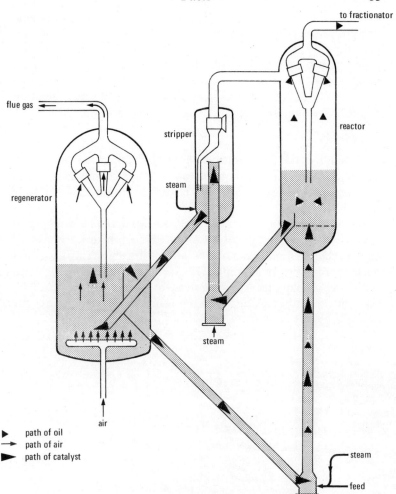

Fig. 2.17 Fluid bed catalytic cracking plant

a second stage reactor to promote further cracking. Using this two-stage technique the output of gasoline is increased and the production of gas and petroleum coke minimized.

Work has recently been carried out in the research laboratories of the Esso Petroleum Co. in which high energy radiation has been used for cracking oil feedstocks instead of thermal or catalytic processes. The radiation, it is claimed, excites individual molecules and strips electrons from them to produce ions. Thus the radiation promotes a supply of free radicals at room temperature which act as initiators for hydrocarbon cracking. The yield is small compared with the amount of energy used but it seems possible that radiation might be used in the future to assist cracking operations.

(b) Reforming

Reforming is used to improve the anti-knock characteristics of gasolines by modifying the structure of their components. As in the case of cracking this can either be carried out thermally or using a catalyst.

Thermal reforming. This resembles thermal cracking in that it involves the breaking down of saturated chain hydrocarbons into shorter chain paraffins and olefins, and the fission of ring structures such as aromatics and naphthenes. As thermal reforming is carried out using gasoline feedstocks instead of the heavier oil distillates used in thermal cracking, higher pressures and temperatures are

A catalytic cracker with a capacity of 33 000 barrels a day used for producing high octane gasoline at the Shell Cardon refinery in Venezuela

necessary. This is because the shorter chain compounds do not rupture as easily as the longer chain hydrocarbons. Under these severe conditions the reforming process must be carefully controlled otherwise large quantities of gas would be formed, lowering the output of gasoline. For the same reason the hot reformed product is rapidly cooled by a stream of cold oil ('quench oil').

The petroleum feedstock is raised to a temperature of 550°C in a heater and maintained at this temperature for the required period of time by passing through a 'soaking' section. The pressure during the thermal reforming is maintained at 8·5–9 MN/m^2 (85–90 atmospheres) and is then rapidly reduced to 0·7–1·4 MN/m^2 (7–14 atmospheres) by means of a reducing valve. After quenching the reformed product as described above, it is passed into a fractionating column where the gasoline reformate is stripped of condensed residue and refinery gas.

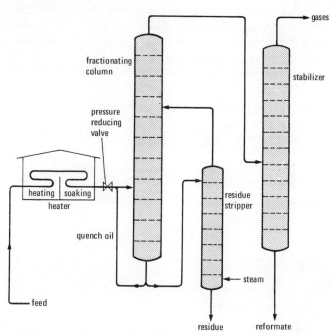

Fig. 2.18   Thermal reforming

Fuels, Explosives and Dyestuffs

During World War II the 'Avaro' (*aviation aromatics*) process was developed at the Shell refinery at Curaçao in the Dutch West Indies in order to boost the production of high octane aircraft fuel. This was a double thermal reforming technique in which a high boiling reformed gasoline was re-reformed in the presence of short chain hydrocarbons to give a high yield of aromatics.

Catalytic reforming. Just as the introduction of catalytic cracking methods improved the quality and throughput of the product, so the use of catalytic reforming has resulted in the production of increased yields of high octane premium grade gasolines. Instead of the alumina/silica type catalysts used for cracking, reforming catalysts usually comprise platinum (up to 0·75% by weight) supported upon an alumina base (carrier). Reforming with platinum catalysts of this type is termed 'platforming' and the resulting reformed product is termed a 'platformate'.

The temperature at which platforming is carried out is around 500°C and the pressure varies from 1–5 MN/m^2 (10–50 atmospheres). The reforming process is complex but includes three main types of molecular rearrangement:

(*a*) Dehydrogenation of naphthenes to form aromatics

e.g. cyclo-hexane $\xrightarrow[\text{500°C 5 MN/m}^2 \text{ (50 atm.)}]{\text{dehydrogenation + Pt}}$ benzene

(*b*) Rearrangement of cyclic structures with side chains to form isomeric naphthenes

e.g. methyl cyclo-pentane $\xrightarrow[\text{500°C 5 MN/m}^2 \text{ (50 atm.)}]{\text{Pt catalyst}}$ cyclo-hexane

(*c*) Saturation of olefins and rearrangement of resulting straight chain paraffins to branched chain isomers

$CH_3(CH_2)_6CH_3$ $\xrightarrow[\text{500°C 5 MN/m}^2 \text{ (50 atm.)}]{\text{Pt catalyst}}$ $CH_3\underset{C_4H_9}{\overset{C_2H_5}{\underset{|}{\overset{|}{C}}}}H$ 3-methyl heptane

n-octane

The type of reformation occurring in (b) and (c) is termed isomerization.

The gasoline feedstock is vaporized and and passed in succession through a battery of three reactors containing the catalyst packed around a central perforated tube. The hot vapour which is at high pressure passes through the catalyst and is removed via the central take-off tube. Reheat furnaces are inserted between the reactors to keep the vapour temperature at about 500°C. Finally the platformate is fractionated to give a series of high-grade motor and aircraft fuels. The reforming process is carried out in an atmosphere of hydrogen which is produced within the plant in the dehydrogenation reaction mentioned in (a) above. This hydrogen not only serves to saturate any olefins formed but removes sulphur as hydrogen sulphide. The volume of hydrogen produced is in excess of requirements and is used elsewhere in the refinery for hydrogenation processes or the removal of sulphur from other products by hydro-desulphurization.

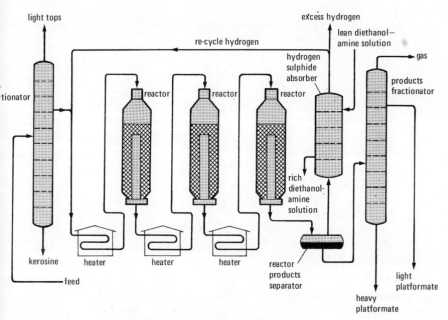

Fig. 2.19   The platforming process

(c) Polymerization

Thermal and catalytic polymerization techniques are also used to synthesize gasolines with high anti-knock properties from the short chain unsaturated olefins which are present in the gas produced from cracking reactions. Unlike the processes discussed above, polymerization involves the building up of more complex hydrocarbons from simple ones.

Thus iso-octane can be synthesized from iso-butylene using a 'hot acid' process involving the use of 70% sulphuric acid; other motor gasolines are prepared from a mixed feedstock containing both C_3 and C_4 olefins and paraffins using a phosphoric acid catalyst.

$$(a) \quad \underset{\text{iso-butylene}}{H_2C=\underset{\underset{CH_3}{|}}{\overset{\overset{CH_3}{|}}{C}}} + \underset{}{H_2C=\underset{\underset{CH_3}{|}}{\overset{\overset{CH_3}{|}}{C}}} \xrightarrow[95°C]{+H_2SO_4} \underset{\text{di-iso-butylene}}{H_3C-\underset{\underset{CH_3}{|}}{\overset{\overset{CH_3}{|}}{C}}-CH_2-\underset{\underset{CH_2}{\|}}{\overset{\overset{CH_3}{|}}{C}}}$$

$$(b) \quad \text{di-iso-butylene} \longrightarrow \text{octylene polymers} \xrightarrow[\text{+ catalyst}]{\text{hydrogenation}} \underset{\text{iso-octane}}{H\underset{\underset{CH_3}{|}}{\overset{\overset{CH_3}{|}}{C}}-(CH_2)_4CH_3}$$

Treating Processes

The final stage in the manufacture of refinery end-products involves the removal of undesirable ingredients, using processes known collectively as 'treating'. Impurities may be present in relatively small amounts (contaminants), as in the case of sulphur compounds, or in more substantial quantities, such as the wax in lubricating oils. Some of the types of treating process in current use can be illustrated by considering four types of product: gasoline, diesel fuel, lubricating oil and bitumen.

Gasoline may contain sulphur as hydrogen sulphide or in the form of thio-alcohols (mercaptans). It is important to remove sulphur compounds from hydrocarbons used as fuel not only because of their objectionable smell, but because of the sulphur dioxide formed on combustion which, in the presence of moisture, is severely corrosive to engine parts and also presents a serious atmospheric pollution problem.

Small amounts of hydrogen sulphide in gasoline are removed by countercurrent washing with a strongly alkaline solution, but where the proportion of contaminant is higher a regenerative process is used to enable recovery and re-use of the absorption material. Two common regenerative processes use potassium phosphate solution (Shell phosphate process) and diethanolamine ('Girbotol' process) as absorbants.

Mercaptans can also be removed from gasoline using solutions of caustic potash or caustic soda, the absorbent power of the alkali being greatly increased by the addition of organic solutizers such as cresylic acids. Regeneration of the spent alkali can be carried out by conversion of the mercaptans to disulphides by air-blowing or by steam-stripping.

Where the mercaptan content of a gasoline is low it is usual to convert it into an odourless disulphide rather than remove it. This is termed 'sweetening'.

$$\underset{\text{mercaptan}}{R\text{—}SH\ HS\text{—}R} \xrightarrow{\text{oxidation in 'doctor' plant}} \underset{\text{disulphide}}{R\text{—}S\text{—}S\text{—}R} + H_2O$$

Originally sweetening was carried out by a 'doctor' process in which conversion of the mercaptan into a disulphide was effected using reagents such as litharge in caustic soda, and copper chloride solution. Great care had to be taken in using copper compounds that none was retained in the gasoline, as copper catalyses the oxidation of lubricating oil and this would interfere with cylinder lubrication. More recently sweetening processes have been introduced which involve passing an air blast through a solutizer solution. The mercaptans are thus converted into disulphides which dissolve in the gasoline. This process is now extensively used.

Gasolines are liable to form resinous gums on standing, due mainly to the polymerization of unsaturated hydrocarbons. This can be prevented by the addition of chemical inhibitors. Gum forming dienes are also removed by selective hydrogenation.

The removal of sulphur from heavier fuel oils is usually carried out by hydro-desulphurization. The feedstock is catalytically hydrogenated under a pressure of 4–6 MN/m^2 (40–60 atmospheres) and at about 360°C. The sulphur is converted into hydrogen sulphide which is stripped from the desulphurized oil. Any unchanged hydrogen is recycled through the plant and regeneration of the catalyst when necessary is carried out by passing a steam/air blast through the reactor bed.

Lubricating oils ('Lube oils') produced by vacuum distillation of suitable heavy oil feedstocks are treated to remove paraffin wax, sulphur compounds, unstable gum-forming substances, and traces of coloured impurities.

Dewaxing is necessary to prevent lubricating oils congealing at low temperatures and to improve their performance as lubricants. Early methods involved simply cooling the oil to a low temperature and then using filter presses to remove the solidified wax. This was not wholly satisfactory, however, as it was very difficult to filter the viscous cold oil rapidly and the process was not continuous. To remove both these objections, continuous filtration techniques were worked out using a 'thinning' solvent for the oil, such as benzene.

The modern solvent dewaxing process involves the solution of the oil (waxy raffinate) in a warm mixed solvent, such as benzene and methyl ethyl ketone (MEK), the latter acting as a wax precipitant. On cooling the solution to $-20°C$ wax separates out and the resulting slurry is passed to revolving drum filters by a worm screw. The surface of the filter drum is covered with tiny slits and these are covered by a filter cloth. Vacuum is applied to the interior of the drum so that as it revolves and the surface dips below the wax slurry the oil is drawn into it leaving a layer of wax on the surface which is continuously removed by a scraper. Apart from the traditional applications of paraffin wax in the manufacture of candles and greaseproof cartons it is also used for 'plucking' broiler fowls and as a cracking feedstock for the production of detergent intermediates.

After dewaxing, the oil is desulphurized by treatment with concentrated sulphuric acid at a temperature between 35–65°C, oleum being used for the manufacture of high-grade medicinal oils and transformer cooling oils. This is followed by clay treatment in which the acidified oil is passed countercurrent through a slurry of Fuller's earth or some similar material at a temperature of 100–180°C and then filtered. This removes unstable gum-forming compounds and colouring matter along with any remaining acid.

The viscosity index of a lubricating oil (a measure of the change

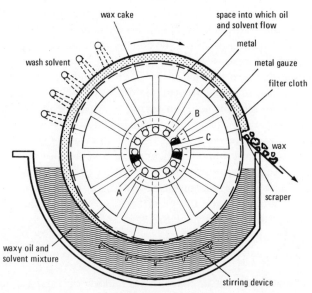

Fig. 2.20   Rotary filter

of viscosity with temperature) can be improved to give 'visco-static' qualities by the addition of high molecular weight polymers such as poly-iso-butene (polybutenes). Other common additives are detergents to suspend sludge in the oil, oxidation inhibitors such as phenyl-α-naphthylamine, and 'extreme pressure' additives such as organo-sulphur compounds to enable satisfactory lubrication under the extreme conditions found with modern hypoid gears. Dispersion of metallic soaps such as calcium or aluminium stearates in lubricating oils are called greases and are normally used for lubricating exposed bearings and other moving parts.

The solid or semi-solid residue remaining after the distillation of crude oil is termed bitumen. The softening point of bitumen is an important characteristic which depends to a certain extent upon the conditions under which distillation was carried out. The softening point can be raised by blowing air through the hot bitumen giving a rubbery tough material ('blown bitumen') very suitable for road surfacing and the manufacture of roofing felts. Bituminous paints and solutions are produced by dissolving bitumen in kerosine, gas-oil or creosote ('cut-back' bitumen). Products of this type are widely used for waterproofing walls and roofs and for protecting pipes and electrical cables against dampness.

Bitumen can also be used in the form of an emulsion in water. The emulsification is usually carried out by slowly adding the hot bitumen to an alkaline solution stirred by a paddle mixing device, or by passing the water and bitumen through a colloid mill. On application the emulsion breaks down allowing the water to evaporate, leaving a bitumen film behind. A novel use for films of this type has been to form a temporary coating over seed beds in dry or dusty regions to retain moisture in the soil until germination is sufficiently advanced for the plant to survive. This technique has been successfully used in the USA.

GAS MAKING FROM LIQUID FUELS

Oil has an advantage over coal as a gas making raw material in that it has a lower carbon/hydrogen ratio which is much nearer to that required in the end product. In addition oil or hydrocarbon gases are much easier to handle and transport than solid fuels and lend themselves more readily to continuous processes. Also the absence of ammonia and hydrocyanic acid simplifies purification. These factors, together with the greatly increased oil refining capacity of this country and the excellent prospects for substantial supplies of natural gas have led to a dramatic swing away from gas making processes using coal. Indeed the traditional method of making town gas by coal carbonization is rapidly becoming obsolete in Britain and has been largely replaced by processes relying upon hydrocarbon feedstocks. The rapidity of this change is reflected in the fact that in 1960 90% of all town gas was still based upon coal, while by 1968 this figure had fallen to 33%. It must be remembered, however, that the advent of North Sea gas has made the oil-based processes rather short-lived. Two principal types of gas are currently produced from naphtha or natural gas in this country. 'Lean gas' (13 MJ/m^3 (350 Btu/ft^3)) is made by catalytic reforming of a mixed steam/hydrocarbon feedstock. 'Rich gas' (24·2 MJ/m^3 (650 Btu/ft^3)) is either produced by catalytic reforming of naphtha with a small proportion of steam or non-catalytic treatment of naphtha with hydrogen. The lean gas is either used to adjust the calorific value of natural gas or rich gas down to 18·63 MJ/m^3 (500 Btu/ft^3) or, after removal of carbon monoxide (as CO_2), as a hydrogen source for the non-catalytic rich gas process.

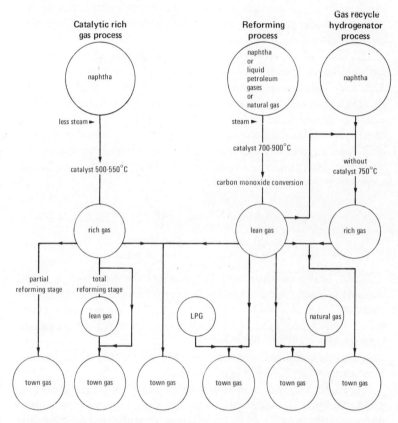

Fig. 2.21 Flow diagram illustrating the ways in which the three basic processes can be combined to produce town gas

The original gas making plants for thermally cracking oil feedstocks used steam without the use of a catalyst. Gas produced by this method was a satisfactory high flame speed substitute for coal gas but an appreciable quantity of carbon black was produced simultaneously at the expense of gas yield—only about half the potential thermal value of the oil being realized. Although thermal cracking processes have now generally been replaced by the more efficient catalytic cracking processes, the market value of carbon black is still high enough to justify the operation of a few of the older plants. An example is the Jones carbon black plant which was built at Gloucester immediately after World War II.

The catalytic type of reactor enables liquid and gaseous hydrocarbons to be cracked without the simultaneous production of appreciable quantities of tarry matter or carbon black. This results in a high gas yield. Catalysts containing lime or nickel as the active agent are most often used, nickel being preferred for the lighter distillates and lime for the heavy oils. These have a dual effect, not only ensuring efficient thermal cracking of the hydrocarbon molecules but also catalyzing the reaction between the latter and steam to produce a water gas mixture.

$$C_aH_b + aH_2O \xrightarrow{\text{lime or nickel catalyst}} aCO\uparrow + \left(a + \frac{b}{2}\right) H_2\uparrow$$

e.g.
$$\underset{\text{ethane}}{C_2H_6} + \underset{\text{steam}}{2H_2O} \longrightarrow \underset{\text{water gas mixture}}{2CO\uparrow + 5H_2\uparrow}$$

When heavy oil distillates are used as the feedstock a small amount of carbon and tarry matter is inevitably deposited on the catalyst. Since this would rapidly poison the catalyst bed a cyclic process has to be used. This enables a blast of air to be passed through the plant to purge the catalyst in alternation with the gas making phase. This is reminiscent of the alternating steam and air blast used in the old water gas process.

Since light oil distillates and refinery gas usually produce negligible carbon deposits, plants using feedstocks of this type can be operated continuously or as cyclic units. Probably the most important advantage of the continuous gas making plants over the cyclic types is that they may be operated at pressures of up to 4 MN/m^2 (40 atmospheres).

Care also has to be taken that fuel gases produced from petroleum by catalytic cracking processes have sulphur-containing substances such as hydrogen sulphide, thiophen and carbon oxysulphide completely eliminated. These materials together with nitrogenous gums and unstable polymerizing hydrocarbons would not only offend the customer but rapidly poison any catalysts used in processing. It is also important that aromatic hydrocarbons be excluded to avoid producing a sooty flame.

Cyclic Processes

As mentioned above, a regenerating air blast is used in cyclic processes to remove carbon and tar deposits. This is usually in the opposite direction to the 'make' blast (i.e. 'counterflow') in the case of heavy oil feedstocks being cracked with a lime catalyst. In the cracking of light oil distillates, naphthas, and hydrocarbon gases using a nickel catalyst, however, both regenerating and make blasts are made in the same direction (i.e. 'uniflow').

The Segas plant consists of three vessels, the steam preheater, the catalyst bed, and the air preheater. Steam is blown through the preheater and carries a spray of oil down through the catalyst bed and out via the air preheater. At the end of the make cycle, air is blown counterflow through the air preheater. The hot air burns up

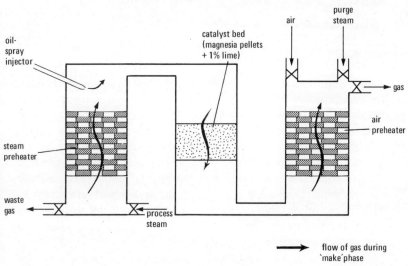

Fig. 2.22 Segas plant

Onia-Gegi gas reforming plant at the North Western Gas Board's Manchester (Partington) works

the carbon deposited on the catalyst bed and then passes through the steam preheater which it reheats. After purging with steam the plant is then ready for the make blast once more. Town gas of 18·63 MJ/m^3 (500 Btu/ft^3) can be produced directly by this means.

The Onia-Gegi process (Office National Industriel de l'Azote and Gaz à l'Eau et Gaz Industriel) which was developed in France is similar to the Segas plant, except that a nickel catalyst is used and the feedstock is sprayed directly on the steam preheater surface. When light oil or refinery gas is used as feedstock the air preheater can be omitted as carbon and tar do not collect on the catalyst bed and the air counterblast is not required. The heating of the steam preheater in this 'two vessel' type of Onia-Gegi plant is effected by burning some of the feedstock during the regenerating blast.

Fuels

There are several other cyclic catalytic reforming processes which are variations upon those mentioned above. The best known of these are the 'P9' and 'P3' processes designed by Gaz de France and the 'SSC' plant developed by the Statione Sperimentale per i Combustibili of Milan.

Continuous Processes

In the continuous type of catalytic reforming process, steam is mixed with a desulphurized hydrocarbon feedstock at a pressure of up to 40 atmospheres and temperature of 700–950°C. The resulting hot gas contains mainly hydrogen, oxides of carbon, methane and unchanged steam. The proportion of carbon dioxide to carbon monoxide can be reduced by raising the temperature, which also has

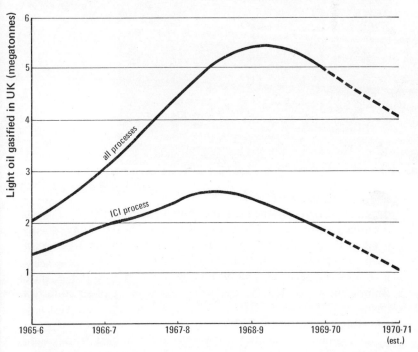

Fig. 2.23   Contribution of ICI naphtha reforming process to oil gasification, 1965–1970

Benfield absorption towers for the removal of CO_2 from town gas produced by reforming naphtha

the effect of reducing the methane content. Raising the pressure on the other hand increases the methane content. The reforming process is effected by heating the oil-gas/steam mixture in the presence of a catalyst in tubes made of nickel chromium alloy steel suspended in a gas-fired furnace. This technique, pioneered by ICI at their Billingham works in 1936, has been used in many other countries including India, Spain, Germany, Japan, France, Austria, Australia and Holland. During the 1960's there was a rapid increase in naphtha reforming plant in the UK and by 1965 over 100 installations were operating with a total throughout of some 500 tonnes of naphtha feedstock an hour.

Fuels

The desulphurization of the naphtha feedstock for the ICI process is carried out in two stages. The first stage involves introducing a small quantity of hydrogen-rich gas and passing it over a zinc oxide catalyst at 400–450°C to remove 'soft' sulphur compounds (mercaptans, hydrogen sulphide and carbon disulphide). In the second stage a molybdenum/cobalt catalyst is used to remove the 'hard' sulphur compounds such as thiophen. These are converted to hydrogen sulphide which is in turn removed by repassing the gas over zinc oxide. The zinc oxide has a life of about four years and can be retained until it contains about 18% by weight of sulphur.

$$\underset{\text{thiophen}}{\begin{array}{c} HC\!\!-\!\!-\!\!-\!\!CH \\ \| \quad\quad \| \\ HC\diagdown_{\!\!S\!\!}\diagup CH \end{array}} \xrightarrow[+H_2]{Mo/Co\ catalyst} H_2S + \text{paraffin hydrocarbons}$$

The reactor tubes, constructed of chrome-nickel steel, are 8·5 m (28 ft) high with an inside diameter of 127 mm (5 in). They are placed vertically in rows in the reformer furnace which is top-heated by oil burners. The tubes are charged with a special catalyst consisting of nickel oxide cemented to a ceramic base (ICI 46-1). Before starting up the plant the nickel oxide is reduced to nickel using a blast of hot hydrogen-rich gas.

Superheated steam, which is produced using the hot flue gas from the reformer furnace, is mixed with the desulphurized naphtha feedstock and passed into the reactor tubes together with a quantity of recycled hydrogen. The 'make' gas leaves the reactor tubes at a temperature of about 700°C and is cooled by passing through heat exchangers designed to preheat the water used for steam raising. After removal of carbon dioxide using hot potassium carbonate solution the calorific value of the gas is adjusted using natural or refinery gas, or other rich gas sources. The enriched gas is then refrigerated and dried. As gases of this type are odourless a chemical is added to produce an artificial smell resembling that of coal gas. The most commonly used odoriferous material of this type is tetrahydrothiophen, about 1 kg (2·2 lb) being added to every 63 000 m^3 ($2\frac{1}{4}$ million ft^3) of gas.

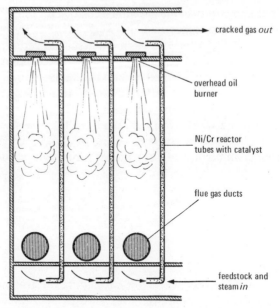

Fig. 2.24 Section of reformer furnace, ICI continuous naphtha reforming plant

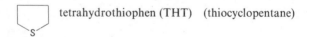 tetrahydrothiophen (THT) (thiocyclopentane)

Ethylene	24%	
Propylene	17%	
C_4 products	10%	
Fuel gas (propane 19% and lighter)		Cracking temperature 800°–900°C
Gasoline	25%	steam: feedstock ratio 1:2
Fuel oil	3%	
Loss	2%	

Yield pattern of steam cracked naphtha feedstock (after Waddams)

The ICI reformer can be modified to produce a wide range of product gases by altering the operating conditions. In this way the plant can be used for the production of hydrogen, ammonia, synthesis gas, town gas and lean gas. Two reformers recently erected at the ICI plant at Billingham have been designed to produce 300 000 tonnes of ammonia or synthesis gas annually. Each plant will process an estimated 500 tonnes of naphtha daily and the erection of even larger plants has been considered.

The UK Gas Council's 'Catalytic Rich Gas' process enables light oil distillate to be used for the production of rich gas. After processing the oil distillate in a continuous catalytic reforming plant of the ICI type the gas stream is split, part being processed in a hydrogenator. Here the gas is enriched by adding a spray of light distillate and then heated under pressure to about 600°C. Under these conditions pyrolysis occurs and the lean gas becomes enriched with methane and ethane. After cooling and passing through a benzole recovery plant the enriched gas is then remixed with the lean fraction.

The success of continuous catalytic reforming plant of the tubular ICI type has stimulated research into this method of gas making using natural gas and oil feedstocks. As a result a number of new processes have made their appearance in various countries, based upon a tubular catalytic reforming unit which has been modified to produce a particular type of gas from locally available gaseous or liquid hydrocarbons.

In the case of continuous autothermic gas making processes the heat required for the reforming process is obtained by burning part of the feedstock. A wide range of hydrocarbon fuels can be cracked in this way, from heavy fuel oils and tar to gases. The feedstock is mixed with air, oxygen, or a mixture of the two before cracking in the reaction chamber. Usually the cracking process is thermal—no catalyst being required. A rich gas is usually produced with a calorific value in excess of 18.63 MJ/m^3 (500 Btu/ft^3), together with the formation of some tar or carbon depending on whether the feedstock is liquid or gaseous. The Shell and Texaco processes have recently been developed to produce synthesis gas in this way by partial oxidation of liquid or gaseous hydrocarbon fuels.

The operating temperature of the cracking chamber is between 1 100°C and 1 500°C at a pressure which may be as high as 4 MN/m^2 (40 atmospheres).

A novel oil hydrogenation process has recently been successfully

ICI continuous steam naphtha/natural gas reforming plant at the Bromley-by-Bow works of the North Thames Gas Board

pioneered by the Gas Council Research Station at Solihull. This uses a bed of crushed coke which is fluidized by a mixed blast of the hydrogenating gas and vaporized oil feed. Using a pressure of 5 MN/m^2 (50 atmospheres) and a feedstock of Middle East crude, a rich gas suitable for a high flame speed town gas is produced.

Experimental work has also been carried out using fluidized sand beds. Sand of grain size 0·4 to 1·2 mm is heated and flows from a bunker into the reactor where it is fluidized by a mixture of superheated steam and hydrocarbon feedstock (anything from ethane to heavy fuel oil). The resulting cracked gas is passed through condensers and an electro-precipitator, while the sand is regenerated by burning off the coke film, thus reheating it. The quantity of sand required is of the order of ten to fifteen times that of the weight of the feedstock. Residence time for the feedstock is only a second or so.

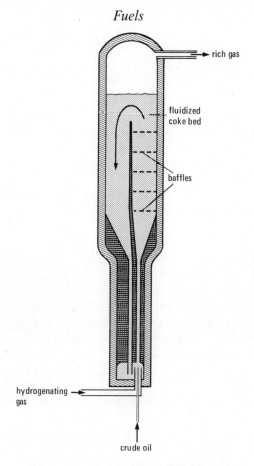

Fig. 2.25   Gas Council crude oil hydrogenator

A mixture of hydrocarbon gases is produced with a high yield of ethylene and methane.

SPECIAL USES OF LIQUID FUELS

The liquid hydrocarbon fuels derived from crude petroleum are of great economic importance. In addition to their widespread use as fuels for internal combustion engines they are in great demand

84   *Fuels, Explosives and Dyestuffs*

for industrial and marine steam-raising and heating plant, and the use of oil-firing for domestic central heating is increasing rapidly. Also many of the petroleum hydrocarbons can be used as rocket propellants.

The most volatile liquid fuels are the liquefied petroleum gases (LPG) which are stored in steel bottles and have many domestic and industrial uses. They are removed from gasoline by distillation under pressure (rectification), a process termed stabilization. Liquid propane is marketed as 'Propagas', and liquid butane as 'Butagas' or 'Calor Gas'.

The colourless volatile fuels boiling between 30°C and 200°C are called *gasolines*, for which an enormous demand was created by the invention of the spark ignition internal combustion (I.C.) engine.

Self ignition of a gasoline/air mixture in the cylinder of an internal combustion engine due to the heat of compression is termed 'knocking', a reference to the characteristic sound emitted. The likelihood of knocking taking place greatly increases as the amount of compression permitted in the cylinder (the compression ratio) increases. Since the thermal efficiency of the engine rises with the compression ratio, much research has gone into the formulation of gasoline fuels with 'anti-knock' characteristics. It was found that high octane fuel containing aromatics and branched chain hydrocarbons had good anti-knock properties which could be markedly improved by the addition of a small quantity (about 1 cm^3/litre) of tetraethyl lead (TEL). The lead oxides formed by combustion of the latter are removed by the addition to the gasoline of a 'scavenger' such as ethylene dibromide. TEL is usually prepared by heating ethyl chloride gas with a lead/sodium alloy.

$$4C_2H_5\text{---}Cl + 4Na/Pb \xrightarrow{\text{autoclave}} (C_2H_5)_4\text{---}Pb + 3Pb + 4NaCl$$

  ethyl   alloy   tetraethyl lead
  chloride   (90% lead)   (TEL)

Other gasoline additives to improve performance are anti-oxidants, such as 2,6-di-tert-butyl-4-methylphenol, and ignition control additives (ICA). Anti-oxidants prevent the formation of gummy deposits in the fuel on standing due to polymerization of olefin peroxides, and ICA suppresses pre-ignition of the fuel due to glowing deposits on a sparking plug or a hot spot on the cylinder wall.

$$(CH_3)_3C\text{-}\underset{\underset{CH_3}{|}}{\overset{\overset{OH}{|}}{\bigcirc}}\text{-}C(CH_3)_3$$

2,6-di-tert-butyl-4-methylphenol

Kerosine (paraffin oil) is less volatile than gasoline and represents a distillate fraction boiling between 140°C and 300°C. Originally of greater importance than gasoline, it was first used as an illuminant in oil lamps. Although eclipsed as a product by the invention of the spark ignition I.C. engine, the demand for kerosine has dramatically increased in the last 20 years due to its use as jet aircraft and tractor fuel, and for space heating. Kerosine used for wick burning must be free from aromatics to avoid smoking. Extraction of the aromatics is carried out by countercurrent washing with liquid sulphur dioxide (Edeleanu process). The inclusion of aromatics in kerosine used as tractor fuel (tractor vaporizing oil) is, however, an advantage as it raises its octane number.

Gas-oils are relatively unimportant as they represent a very small distillate fraction. They are brownish oils boiling just above the kerosines and are used to enrich water gas and as cracking feedstocks for the production of premium gasolines.

Diesel fuels are heavy gas-oils used in diesel engines which rely upon compression to ignite the fuel/air cylinder mixture. The smaller fast running diesel engines used for buses and trucks run on a light diesel oil with a cetane number of 45–50. The large slower running types used for electricity generation and marine engines use a viscous fuel with a cetane number of 25–35 containing a proportion of 'residual fuel'. This residue produces a carbon deposit known as 'coke' when the fuel is burnt and has to be limited to prevent a build-up of this deposit in the engine.

Fuel oils are very dark viscous liquids produced as residues from the distillation of crude oils with a high asphaltic content. They are used as under-boiler fuels to raise steam for ships, locomotives and industrial purposes. Before burning, the viscosity of the fuel is reduced by pre-heating and owing to its low volatility the fuel is 'atomized' or broken into an exceedingly fine spray before ignition. Fuels of this type have an ash content, but this is only about 0·1 % of the total weight of the fuel.

Aviation gasoline and kerosine are used as rocket propellants. They rank high amongst rocket fuels in current use because of their availability, low cost, and desirable physical properties such as low freezing point, high bulk density and high jet velocity. Specially prepared jet fuels are coded JP. Fuels JP1-3 are now obsolete but JP4–6 are still in use. The US Air Force uses a special hydrocarbon rocket propellant known by the code name RPI.

There are two types of chemical rocket engine—one burning liquid propellant and the other solid propellant. In a liquid-fuelled rocket, fuel and oxidizer are carried in separate tanks and burned in the combustion chamber of the rocket engine. The most common fuels are hydrocarbons such as kerosine, and this type of fuel has been used in rocket boosters such as Atlas, Delta and Saturn. The giant Titan 11 vehicle however uses a mixture of unsymmetrical dimethyl hydrazine (UDMH) and hydrazine. The most usual oxidizers are liquid oxygen, nitrogen tetroxide and red fuming nitric acid. There is, however, a move towards the use of higher specific impulse fuels such as liquid hydrogen. Experimental work with liquid hydrogen and liquid fluorine as a propellant combination has produced promising results.

Solid fuel rockets have at the moment only been used for small-scale military and sounding work, although work is still being actively carried out in this field. The largest solid charges at present in use have a diameter of over 6 m (20 ft) and produce around 1·36 Gg (3 million lb) of thrust. In solid fuel rockets the oxidizer and fuel are mixed and bonded to the interior of the motor casing. The shape of the inner surface of the charge is designed to maintain a constant area of burning. One of the drawbacks of solid propellants is the difficulty of controlling the thrust rate.

Synthetic elastomers such as Thiokol have also been successfully used as solid propellant fuels for rockets, being particularly suited for this purpose as they have an unusually high heat of combustion (42–44 MJ/kg (18–19 000 Btu/lb)). The rocket casings are filled

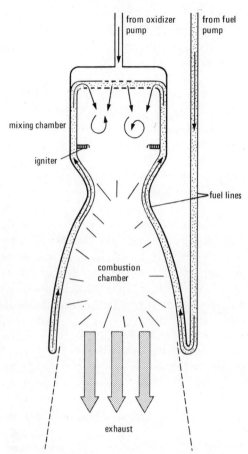

Fig. 2.26 Combustion chamber of liquid fuel rocket

by pouring in a thin paste of depolymerized rubber in mineral oil, to which has been added some sulphur as a vulcanizing agent and an oxidant such as ammonium nitrate. After filling is complete the rocket is heated to re-polymerize the rubber which sets to a solid mass. The Polaris submarine missile is powered with a propellant based upon Thiokol.

There have recently been attempts to produce a single propellant system for rockets in order to simplify their construction. The

Rocket Research Corporation (Washington) has developed a small propellant unit which uses hydrazine fuel (b.p. 113·5°C). A special catalyst (Shell 405) is used to decompose the hydrazine into nitrogen, hydrogen and ammonia.

$$2NH_2 \cdot NH_2 \xrightarrow{\text{Shell catalyst 405}} 2NH_3\uparrow + N_2\uparrow + H_2\uparrow$$
hydrazine

The low thrust and simplicity of construction makes the hydrazine rocket motor a viable proposition for altitude control nozzles on satellites.

An interesting problem has arisen in the design of fuels for supersonic aircraft. Air friction causes the outer skin of an aircraft to heat up, the temperature rising rapidly with the speed. The fuel of the aircraft is used to cool the walls by circulating it through heat exchangers. This not only reduces the skin temperature but heats the fuel, thereby increasing the efficiency of the jet engine. Unfortunately the earlier JP fuels deteriorated at the temperature experienced in supersonic flight, producing deposits which fouled injectors and pipes. The most suitable kerosine-based fuel (JP 6) is thus only suitable up to speeds of Mach 3. Above this speed new fuels have had to be tried and some valuable work has recently been carried out in France by CERCHAR (Centre d'Etudes et Recherches des Charbonnages de France) to produce a fuel ('Carbojet') by the hydrogenation of coal tar. This is carried out using tungsten and molybdenum sulphide catalysts which are not poisoned by organic sulphides. Heterocyclic components of the tar containing nitrogen, sulphur and oxygen atoms are broken down during the hydrogenation and controlled cracking occurs which eliminates aromatics with more than three rings.

Although expensive and limited by supplies of coal tar, 'Carbojet' is thermostable even at temperatures as high as 274°C, is non-corrosive and has a freezing point below −60°C. It contains several saturated multi-ring structures including di-cyclo-hexyl, perhydrofluorene and decalin.

decalin

di-cyclo-hexyl

perhydrofluorene

Fuels

A novel development in the last few years has been the manufacture of protein from hydrocarbon sources such as gas-oil and methane. Experiments carried out by French workers in the 1950's showed that yeasts of the *Candida* species could metabolize gas-oil. As a result a research department was set up at Lavera (France) in 1960 to assess the commercial possibilities of the process and British Petroleum built the first large-scale plant there in 1968. A similar plant is projected for the UK at Grangemouth.

The BP process uses either a gas-oil feedstock of purified straight chain paraffins. The selected yeast strains are then dispersed in an oil-in-water emulsion containing ammonium salts, and agitated with sterile air.

The use of methane for protein production rather than gas-oil has a number of advantages including cheapness and the production of a 'clean' paste of yeast which does not have to be purified by extraction with a solvent. Shell are experimenting with nitrogen fixing bacteria which would metabolize methane without the need for the addition of ammonium salts. These convert the methane into protein by a process which is analogous to the formation of carbohydrate from carbon dioxide by plants. Instead of the energy for the process being derived from sunlight, however, as in photosynthesis, the methane itself supplies the energy necessary for its conversion into protein. A number of pure cultures of highly efficient methane-oxidizing bacteria have been isolated from natural sources by the Shell scientists, and it is estimated that 5 tonnes of edible protein can be obtained from 28 000 m^3 (1 million ft^3) of natural gas.

The cost of hydrocarbon derived protein is expected to be about £50 per tonne, which is competitive with natural proteinaceous materials such as fishmeal (£65 per tonne) and soyameal (£50 per tonne) and it is expected to be used primarily in cattle and poultry feed.

Initial experiments to determine the feasibility of using gas-oil grown yeast as an animal food have shown that this can form an effective alternative to traditional high protein feeding stuffs based on soya and fishmeal. In the case of pigs and poultry, growth rate, fertility and general health were all found to be normal if methionine was added to the feed. Fears that yeasts produced from oil substrate would contain toxic or carcinogenic substances have also been discounted and consumer trials show that the flavour of pork from yeast fed pigs is indistinguishable from that obtained from control animals.

(3) NATURAL GAS

The exploitation of natural gas in Europe since the end of World War II has been one of the most remarkable developments in the history of fuel technology. In addition to the enormous quantities of gas associated with oil or coal deposits, smaller but valuable quantities of methane are produced by the digestion of sewage.

Although during 1937 the BP Exploration Company discovered two methane deposits at Eskdale and Cousland (Scotland), little further exploratory drilling was carried out until 1953. In that year a productivity team appointed by the Gas Council reported: 'Although there are no appreciable known reserves of natural gas in Great Britain, a discovery of any magnitude would be of immense value to our national economy'. In November of the same year an intensive seismic survey was made in the East Riding of Yorkshire, the geological structure of the underlying rock being determined by the vibrations set up on firing small explosive charges. Unfortunately little gas was found.

In November 1961 the British Government approved a Gas Council scheme for the importation of liquid methane from the Hassi R'Mel field in the Sahara desert, using a converted 5 000 tonne cargo vessel re-named the *Methane Pioneer*. In 1963 two specially designed tankers, *Methane Princess* and *Methane Progress*, were launched, and a contract was signed with Algeria to supply the UK with $2 \cdot 8$ hm^3 (100 million ft^3) of natural gas per day. The liquid methane is transported at a temperature of $-160°C$ in aluminium tanks insulated from the bottom and sides of the ship by balsa wood and glass fibre panels. The 'boil off' of fuel which occurs in transit is used to supplement the ship's boiler fuel.

The cargo is pumped into special storage tanks erected at Canvey Island (Essex) having a total capacity of 22 000 tonnes of methane. From here it is vaporized and pumped into a methane grid which links up with Leeds, Sheffield, Manchester, London and Reading. The grid has recently been extended to accommodate natural gas from the North Sea fields. Feeder mains have been laid from the coastal towns of Easington (Yorks.) and Bacton (Norfolk) to the London-Sheffield trunk line. By 1970 all twelve area gas boards should be receiving their quota of natural gas, with the construction of spurs to Wales, Scotland and the West Country.

Great care has to be taken in the design of liquid gas tanks because of the brittleness of steel at low temperatures. This is thought to have been the cause of the Cleveland (USA) disaster in 1944 when two liquid gas holders collapsed and 5 700 m^3 (1$\frac{1}{4}$ million gallons) of liquid methane ignited, killing or injuring over 500 people.

An ingenious alternative method of liquid gas storage is in suitable underground reservoirs such as disused coal or salt mines. The first underground reservoir to be put into operation in Britain was in a rock salt mine on the north bank of the River Tees which was constructed in 1959 at one-sixth of the cost of building a conventional gas holder of the same capacity. Test borings have also been carried out in the vicinity of Winchester where suitable geological formations have been discovered.

Liquid natural gas may also be conveniently stored in large holes in the ground covered with a thermally insulated roof. The low temperature of the gas ($-160°C$) freezes the soil into a hard non-permeable layer which forms its own seal. Two gas reservoirs of this type have been installed at Canvey Island, each having a capacity of 21 000 tonnes of liquid gas equivalent to some 1 160 TJ (11 million therms).

During the period from 1962–64 the hunt for natural gas in Britain was renewed and an extensive survey was made of the North Sea bed from the English Channel to the Orkneys. Up to 12 seismic survey ships took part in the operation which covered tens of thousands of square kilometres of sea bed. Very large size structures at great depth were located and there was evidence that the large salt beds characteristic of the Dutch gasfields extended well across the 'median line'. This demarcation line separated the British and European territorial areas and was drawn up by the Continental Shelf Act of 1964.

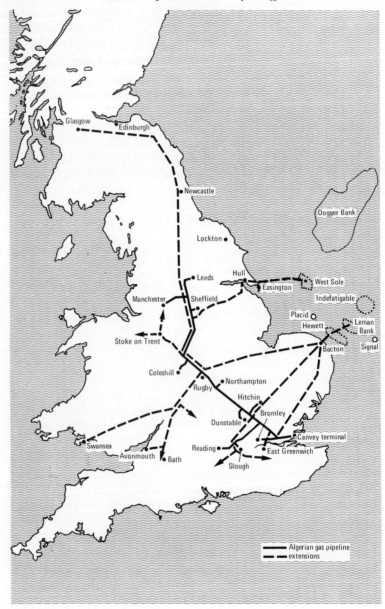

Fig. 2.27 UK methane grid

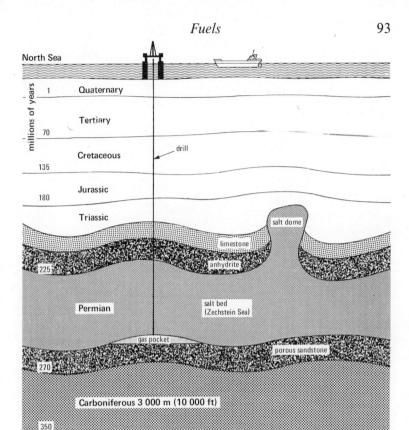

Fig. 2.28 Section through the North Sea bed

After the sea survey was completed concessions were granted by the Government in 1964 to 23 companies, allowing them to drill for natural gas in specified blocks of about 260 km² (100 sq. miles). About one-third of the area (93 000 km² (42 000 sq. miles)) has so far been licensed, the largest operators being Shell/Esso (89 blocks) and Gas Council/Amoco (51 blocks).

Drilling was carried out by various types of off-shore drilling rig. Self-elevating jack-up platforms were used up to a depth of 90 m (300 ft) and semi-submersible rigs for deeper work. The rigs house all drilling personnel and are equipped with helicopter landing pads.

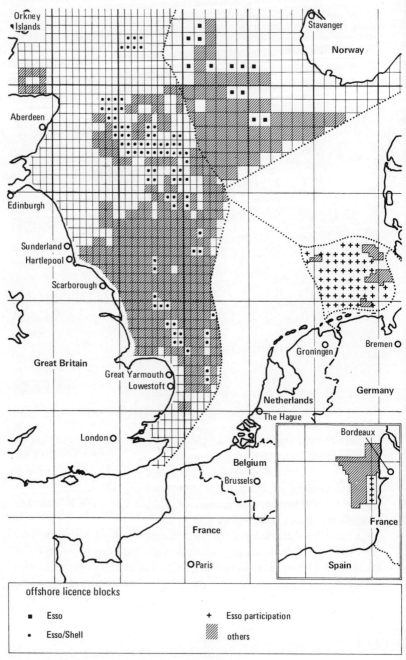

Fig. 2.29

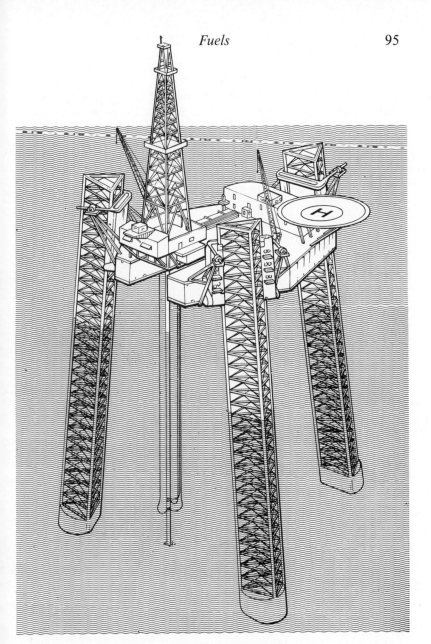

Fig. 2.30 Self-elevating underwater rig

Fuels, Explosives and Dyestuffs

Drilling was commenced in the Dogger Bank area on Boxing Day 1964. The first commercial gas strike was made by the ill-fated British Petroleum rig 'Sea Gem' during October 1965 in the West Sole field about 64 km (40 miles) east of Grimsby. Since then three other major gas-producing areas have been located off the Norfolk coast, the Leman Bank, Indefatigable and Hewett fields.

Between them these four fields have recoverable reserves of around 708 hm^3 (25 billion ft^3). This is equivalent to an average daily output of about 70·8 hm^3 (2 500 million ft^3) over a period of 30 years,

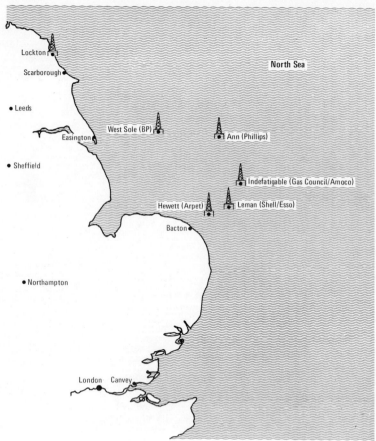

Fig. 2.31 Natural gas strikes in the North Sea and Yorkshire

allowing for a fall-off in production as reserves are depleted. This is two and a half times the present daily send-out of gas in the UK and will therefore lead to a rapid expansion of the industry, which at the moment is growing annually at the rate of about 10%. The first large land-strike was made at Lockton, a small Yorkshire village 19 km (12 miles) west of Scarborough, in May 1966.

The first experimental quantities of gas were brought ashore at Easington on the Yorkshire coast by means of a 80 km (50 mile) undersea pipeline in March 1967. A feeder main was then constructed from Easington to join the existing Canvey-Leeds methane main near Sheffield. A second pipeline from the Indefatigable, Leman and Hewett fields brings gas ashore at Bacton (Norfolk). From here it is pumped to the methane grid which it joins at a point near Rugby.

	Mol. %
N_2	1·4
CO_2	0·6
C_1	94·0
C_2	3·0
C_3	0·6
$i-C_4$	0·1
$n-C_4$	0·1
C_5	0·08
C_6	0·05
C_7	0·03
Benzene	0·04
C_8	0·01
Toluene	0·01
	100·02

calorific value 38 MJ/m^3 (1 020 Btu/ft^3)
specific gravity 0·598

Gas analysis: West Sole field

In 1966 a pilot experiment was carried out to convert the equipment of some 8 000 households on Canvey Island to natural gas. On May 15th, 1967, the two Staffordshire villages of Fradley and Alrewas were the first to be converted to natural gas in a programme

Construction of the 54-km, 76-cm marine pipeline which carries gas from the Leman field in the North Sea. The sea and shore pipelines are being manoeuvred before joining up at Bacton, Norfolk

intended to cover most of the country by 1978. A local newspaper described the scene on the morning of G-day as follows: 'A 150-strong team spread throughout the two villages knocking on doors for permission to turn off gas supplies at the meter. When the signal had been radioed back that all gas supplies were shut off, workmen stationed at half a dozen valves closed them to isolate the villages from the main town gas line. Algerian gas from the Canvey-Leeds pipeline was then forced through the system, expelling the town gas which was fired at a number of "flare offs". Meanwhile the conversion fitters had begun work as soon as the house meters were turned off.'

In some countries the use of natural gas is well established. Almost a third of the total energy requirements of the USA is obtained from this source which exceeds the contribution made by coal. The consumption of natural gas is expected to have risen by 1975 to the prodigious quantity of 538 hm^3 (19 billion ft^3) a year.

The largest natural gas producer in Europe and Asia is Russia with an estimated annual production rate of 340 hm^3 (12 billion ft^3) by 1972. Almost 80% of the fuel at present used in Moscow is natural gas. Russia has also installed a considerable amount of plant for the processing of natural gas, the output of liquid gas in 1965 being 4 million tonnes.

Other large-scale natural gas deposits have been discovered in Canada (Alberta and Ontario), France (Lacq), Holland (Groningen), Italy (Po Valley), Sahara (Hassi R'Mel), Romania and Mexico. Most of the gas from these sources is 'dry', but if it is found associated with oil the latter has to be removed by passing the gas through a separator. An experimental project called 'Operation Gas Buggy' has been carried out in the USA to explore the possibility of recovering natural gas trapped in relatively small isolated pockets. An underground nuclear explosion was used to break up the underground gas-bearing rock formations allowing the gas to filter through to a convenient collection point. Both the test explosions carried out to date have been successful.

Occasionally a high proportion of hydrogen sulphide is present, as at Lacq (15%) and in some of the Alberta deposits (up to 80%). This creates special technical problems both in the purification of the gas and in its recovery. Atomic hydrogen liberated by the action of the hydrogen sulphide gas on iron permeates the steel borehole lining producing a glass-like brittleness which often destroys the piping in a few hours. A specially heat-treated steel alloy containing $1\frac{1}{2}\%$ chromium and 1% aluminium has been used successfully at Lacq.

Removal of hydrogen sulphide from natural gas is usually carried out by absorption in a liquid such as aqueous monoethanolamine, which is run countercurrent to the gas flow. The hydrogen sulphide gas can then be stripped from the absorbent by heating and then oxidized to sulphur. Solutions of zinc acetate, or zinc sulphate in aqueous sodium hydroxide, have also been used to remove hydrogen sulphide, and experimental work has been carried out using organic oxidizing reagents such as methylene blue chloride. Dehydration of the gas is also important to prevent icing and condensation problems. The removal of water can be carried out by passing through beds of silica gel or liquids such as triethylene glycol. Freeze drying has also been tried in the USA. As with the gas produced from reformed petroleum feedstocks, processed natural gas is almost odourless and a small amount of tetrahydrothiophen is added to give it a characteristic 'coal gas' smell.

An unexpected complication arising from the use of tetrahydrothiophen (THT) with North Sea natural gas has been the loss of much of its odour. This has been found to be due to the masking effect of certain constituents in the gas, and has made it necessary to find an alternative odorant compound. After a number of substances had been tried without success, it was decided to analyse the natural gas from the Leman field which has such an unpleasant smell that no additional odorant is necessary. As a result of this a mixture known as Robodorant A is being adopted throughout Britain as a gas odorizing agent. In some cases the same effect will be achieved by mixing the excessively odorous Leman gas with that from other fields.

The first North Sea gas containing sulphur was reported from the Hewett field. Before being used as a fuel this 'sour' gas is treated to reduce the hydrogen sulphide content to the level required by the Gas Council. This is accomplished by first washing the gas with ethanolamine solvent. Steam is then used to remove the dissolved hydrogen sulphide. This is partly burnt off and partly recovered as sulphur.

The potential throughput of the present sour gas plant in Bacton (Norfolk) is 17 hm^3 (600 million ft^3) daily which represents around 12 tonnes of sulphur. The recovery of sulphur by this means is uneconomic but burning it off as sulphur dioxide would cause air pollution problems in the neighbourhood of the plant.

Negotiations over the price at which producers will sell North Sea gas to the Gas Council have been protracted. This is understandable if the enormous sums of money involved are considered. At the price of 1·33p/100 MJ (2·87d per therm) agreed in March 1968 by the Phillips group the total revenue from 708 hm^3 (25 billion ft^3) of gas is £3 000 million. ICI have recently agreed to take £250 million of natural gas over a 15-year period to replace naphtha and fuel oil in the manufacture of ammonia gas.

In countries such as Britain, Germany and Belgium which have extensive coal deposits, considerable quantities of methane can be extracted from coal mines. In the UK something like $8\frac{1}{2}$ hm^3 (300 million ft^3) of methane a day is available from this source, of which only a small fraction is at present being used. Methane drainage plant is being used (1969) in 120 pits, yielding about $8\frac{1}{2}$ hm^3 (300 million ft^3) of gas a week, about half of which is used either as fuel at the colliery itself or sold to neighbouring gas undertakings.

Methane drainage is carried out by boring from the mine roadways into the surrounding strata to a depth of about 12–60 m (40–200 ft). The open end of the borehole is fitted with a 76 mm (3 in) steel tube which is joined to the main shaft pipes, which are 250 mm (10 in) or more in diameter. Although areas being currently worked are those usually drained, methane can be collected from worked-out areas. After passing through a cooling plant to remove moisture, the gas is usually supplied to the local Gas Board as rich gas having a calorific value of around 26 MJ/m^3 (700 Btu/ft^3). On the Continent, mine gas is still used to a limited extent as a motor-car fuel for use in specially adapted engines. The Russians have also attempted to adapt cars to run on liquid methane gas.

The production of methane from sewage (sludge gas) is also growing in importance, and large modern sewage installations are already equipped for this purpose. The sewage sludge which contains finely suspended and dissolved organic matter is 'primed' by adding a quantity of 'activated' sludge, and after aeration by blowing with compressed air is allowed to remain at about 85°C in a digestion tank for three weeks.

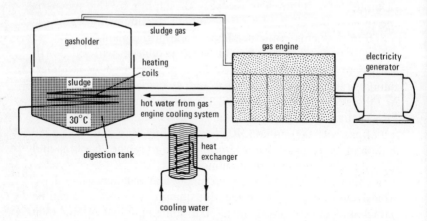

Fig. 2.32   Sludge gas producer

During this time sludge gas is evolved and is collected in a gas holder, being a mixture of some 70% methane and 30% carbon dioxide. Hydrogen sulphide is also present and presents the same problems already encountered with natural gas (see above). Most of the gas is used to generate power in the plant itself, as at the Mogden plant of West Middlesex Main Drainage which produces 42 500 m^3 (1½ million ft^3) a day.

Plant for the production of methane from farmyard waste and manure has also been designed. A small installation on a 36 hectare (90 acre) Gloucestershire farm which was erected in 1954 is used for domestic heating and for powering a 1·12 kW (1½ hp) engine.

Chapter 3
Chemicals from Coal and Petroleum

COAL CHEMICALS

The use of coal as a source of carbon chemicals stems from the discovery of coke as a smelting agent by the Quaker iron-master Abraham Darby during the latter half of the seventeenth century. The first British patent for producing pitch and tar from coal was granted in 1681 to Becher and Serle who envisaged using these novel materials to replace 'Stockholm' wood tar in the preservation of cordage and wood. The process does not appear to have been commercialized for another hundred years, however, when Dixon of Cookfield set up a pioneer coal-carbonizing plant in 1780. At about this time the Earl of Dundonald built a similar plant on his estate at Culross Abbey on the banks of the river Forth near Edinburgh, and founded the British Tar Co. In addition to his interest in coal tar, the scientifically minded Earl also lit the Abbey by gaslight, although he had been anticipated in this respect by William Murdoch.

The production of coal gas on a large scale brought with it the problem of how to dispose of the coal tar formed as a by-product. Although the distillation of coal tar had been carried out in 1818 by Accum, most of it was still used to waterproof cordage, to treat ships' bottoms or was burnt as fuel. When the Naval authorities discontinued the use of tar in the 1830's new outlets for this substance were sought. In 1838 Bethell patented a pressure impregnation process for creosoting railway sleepers—then much in demand—and pitch was first used in 1842 for briquetting coal.

The rapid growth of organic chemistry, however, soon indicated that coal tar was a valuable source of new chemical substances—

benzene, phenol, aniline, naphthalene, toluene, anthracene and the cresols being isolated from it in the space of a few years. 'Gentlemen,' said Hofmann to his students at this time, 'new bodies are floating in the air'. The discovery that set the seal on coal tar as a commercial product was undoubtedly the discovery of aniline dyestuffs by W. H. Perkin in 1856. The coal tar industry in this country was soon to be crippled by competition from Germany, however, and by the turn of the century the price of crude tar had fallen from £3·50 to 35p a tonne. The demand for explosives and other heavy organic chemicals during World War I gave the industry a much needed boost. After the war the development of phenolic resins and the road tar industry, coupled with an increased demand for benzole fuel ('National Benzole'), led to steady growth of the coal tar industry, except for the period of the economic depression in the 1920's. The design of tar distillation plant also improved from about 1920 onwards enabling more effective concentration of the valuable fractions.

Since World War II the importance of coal tar as a source of organics has declined, many of the products traditionally produced from tar distillates being synthesized more cheaply from petroleum

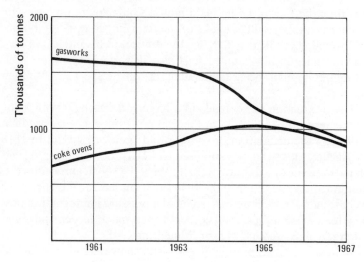

Fig. 3.1 Production of coal tar in the UK

Chemicals from Coal and Petroleum

or other sources. The production of coal tar in Britain has been falling steadily as natural gas and cracked naphtha become increasingly used as sources of town gas.

Some coal chemicals like naphthalene are commercially valuable and in short supply; others such as pyrene, phenanthrene and chrysene are not recovered. Crude tar is usually fractionated by distillation to give a number of tar oils and a residue of pitch. The number of fractions and their content depends upon the type of retort and the coal used for carbonization. Thus tars obtained by fluidized bed techniques at a comparatively low temperature (420°C) have an exceptionally high phenolic content but little benzene or naphthalene. The problem of the tar distiller resembles that of the oil refiner in attempting to maintain the maximum output of high demand products from a raw material of varying composition. It is important to remember that only about 10 per cent of the crude coal tar produced is used as a source of coal chemicals. The major proportion is used as road tar, heavy fuels, pitch and crude creosote.

	(thousands of tonnes)			
	1963	1965	1967	% of total
Refined tar	574	564	580	27
Creosote/pitch fuels	1 008	796	737	34
Creosote oil	430	445	405	7
Pitch	418	463	308	14
Total tar distilled	2 583	2 401	2 152	100

Production of heavy coal tar products in Britain

Most of the tar processing in the UK is carried out by co-operative tar distilling firms such as the giant Midland–Yorkshire Tar Distillers Company, which has an annual throughput of some 700 000 tonnes of coal tar. Some producers still process their own tar. These include the North Thames (Beckton), South Eastern (Greenwich), Northern (Gateshead) and the North Eastern (Hull) Gas Boards, the National Coal Board, two steel companies, and some of the gas-making plants still based on coal.

About 7% of the coal tar distilled currently in the UK is processed in pot stills, but continuous tube stills are most commonly used. In these the tar is heated by passing through a series of cast iron pipes set in a furnace burning gas, oil or pitch fuel. From here the heated tar is passed into a flash chamber where the tar oils vaporize off leaving the pitch residue. The vaporized tar oils then pass up a bubble cap column containing about 40 plates, the fractions being drawn off at different levels. Before distillation the crude tar feedstock is stored for some days in a steam-heated tank and then filtered to remove solids and entrained liquor.

In addition to the 'once-through' type of still (ProAbd), there are a number of continuous stills designed to recycle the feedstock using dual flash chambers (e.g. the Wilton still) or in some cases vacuum distillation techniques (e.g. the Raschig still).

Coal tars are exceedingly complex mixtures of organic compounds, consisting of hydrocarbons, heterocyclics and their alkyl and hydroxy-derivatives, many of these occurring in fused ring systems. A detailed treatment of the isolation and purification of all these substances is not possible in a general account of tar distillation, but it is interesting to examine the components of the main fractions.

LIGHT OIL

(up to 170°C)

1–6%

Contains mainly crude naphtha with some benzene, toluene and xylenes (known in the industrial crude form as benzoles, toluoles and xyloles respectively). These are used mainly as solvents for paints and polishes, although benzene is used in the synthesis of aromatics such as aniline, styrene and phenol. Toluene is in demand for explosives and the manufacture of saccharine and polyurethanes. The light oil fraction is small because most of the low boiling hydrocarbons remain entrained in the coal gas and are removed during the gas-oil washing stage.

MIDDLE OIL

(170–230°C)

6–8%

Known originally as 'carbolic oil', as the lower boiling fraction up to about 200°C contains mainly phenol, cresols, and xylenols. These are used as disinfectants and in the production of resins and plasticizers. Phenol is also used in the

A modern continuous tar still, Wilton design

synthesis of aspirin and nylon. The high boiling fraction is mainly naphthalene which is a useful insect repellent and is used in the manufacture of dyestuffs and resins.

HEAVY OIL

(230–270°C)

8–13%

This fraction contains a high proportion of 'creosote', a blend of high boiling-point tar oils from which the more valuable components, such as anthracene and the tar acids and bases, have been removed. Mixtures of pitch and creosote produced from this fraction are used as liquid coal tar fuels (CTF) for heating furnaces. These require preheating before burning, the appropriate temperature (in °F) being indicated by a number (e.g. CTF 200).

Creosote oils are also widely used for preserving wood and in the formulation of sheep dips and fruit sprays. A plant was set up in 1935 at the ICI Billingham works for the hydrogenation of creosote to produce petrol. This was of great value during World War II by which time annual

Fuels, Explosives and Dyestuffs

production of petrol had reached 150 000 tonnes. The fall in petroleum prices after the war made the process uneconomic however and production ceased in 1956.

ANTHRACENE OIL (GREEN OIL) (270–350°C)

This fraction contains very small quantities of anthracene and phenanthrene, which are removed as a crude coke by cooling and filtering. Anthracene is used in the manufacture of anthraquinone dyestuffs.

50–60%

anthracene   phenanthrene

The other components of the fraction are heavy coal tar fuels (CTF 200–300) and road tar.

PITCH

15%

The pitch residue is used as a fuel, as a binding agent for the production of smokeless fuel briquettes, weather-proofing roofs, and for the manufacture of anti-corrosion metal paints.

Distillation of the benzole obtained from coal gas by gas-oil extraction also yields a number of fractions. These are principally benzene, toluene and xylene, together with appreciable amounts of solvent naphthas (b.p. 160–190°C) which are used as solvents for paint and bitumen and as a source of polymerizable resin bases such as styrene, coumarone and indene.

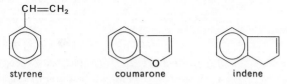

styrene   coumarone   indene

Crude benzole from vertical retorts is less valuable than that obtained from the old horizontal retorts and coke ovens. For this reason it is often only recovered from coal gas when this is necessary to avoid mains condensation, or when its value as a motor fuel exceeds its enrichment value in the gas. In fact, since the abolition in 1962 of the preferential excise duty on benzole of 1·4p/l (1/3d a gallon), its removal from coal gas is only profitable because it is the simplest way of removing dissolved sulphur compounds such as carbon disulphide, mercaptans and thiophen.

	millions of gallons (4.5×10^4 hl)			
	1964	1965	1966	1967
Total crude benzole	114·4	106·1	96·6	90·9
Refined benzole products (a) motor benzole (b) refined benzole (c) toluole (d) xylole (e) coal tar naphtha	30·1 41·7 9·3 4·1 11·1	25·6 40·1 8·8 3·1 10·5	23·8 35·2 8·6 2·5 9·6	17·6 35·0 8·9 2·8 9·2

Production of crude benzole and refined light oils (UK)

The light and middle oils are neutralized with aqueous sodium carbonate and then treated with carbon dioxide and steam in countercurrent reaction towers to regenerate the free 'tar acids' (i.e. phenols, cresols and xylenols) which are then separated by distillation. Naphthalene is extracted from the neutralized middle oil by allowing it to crystallize out in shallow pans at room temperature and then centrifuging. After hot pressing the 'whizzed' naphthalene is pure enough for the manufacture of phthalic anhydride but if a higher standard of purity is required the molten impure product is treated with sulphuric acid. After neutralizing the naphthalene is fractionally distilled.

The creosote fraction of the heavy oil is not further processed in Britain but the anthracene is allowed to crystallize out and is removed by centrifugation. Tar bases are also removed from the light, middle and heavy oil fractions as sulphuric acid extracts, although benzole is a more important source. About 2% of crude tar consists of bases such as pyridine and picolines.

	thousand tonnes		
	1965	1966	1967
Crude naphthalene	81·6	81·1	77·6
Pure naphthalene	3·9	3·6	7·0
Phenol	18·4	18·2	17·3
Cresylic acid	73·1	69·2	67·5
Anthracene	4·0	3·5	3·6
Pyridine bases	6·7	5·1	4·9

Production of coal chemicals (UK)

In addition to the organic by-products of coal carbonization mention must be made of the ammonium sulphate produced by neutralization of the ammoniacal liquor. This is rapidly decreasing in importance due to the large-scale manufacture of ammonia from synthesis gas, natural gas and refinery gases. Before World War II over 90% of the world production of ammonia was based on coal. By 1958 this had fallen to 40% and by 1968 was little more than 20%. In the UK there was a massive changeover from coal-based to oil-based ammonia after 1957.

Attempts have been made to produce acetylene directly from coal using plasma arcs. Theoretically this should produce high yields but at the moment the project is still at the research stage. Another interesting development has been the use of solvents to extract specific chemical fractions from raw coal without carbonization. This has already been mentioned in connection with ashless fuels in the previous section.

PETROLEUM CHEMICALS

Petroleum and natural gas are now firmly established as the preferred sources of hydrocarbon raw materials used in the production of organic chemicals. With an output of about $1\frac{1}{2}$ million tonnes a year (1969) the UK is the second largest producer of chemicals from petroleum in Western Europe. This achievement is all the more spectacular when it is borne in mind that the first plant for the production of chemicals from petroleum did not begin to operate in this country until 1942 and large-scale production only got underway in the early 1950's.

The rapid expansion in the manufacture of petroleum-based chemicals in Britain (800% in the decade 1954–1964) has been mainly due to the demands of the thrusting new polymer industries, especially in the fields of fibres and plastics, and the greater availability of petroleum hydrocarbons due to a massive increase in UK refinery capacity since World War II. Also, countries such as West Germany (the leading producer of petroleum chemicals in Europe) and the UK, whose economies have traditionally been based upon coal and coal chemicals, have been compelled to switch their allegiance to oil because of its competitive price and ease of transformation into other organic compounds. Even if coke oven and coal tar refining capacity could be increased to meet the enormous demand for organics from this source, insufficient quantities of coking coal would be available for carbonization.

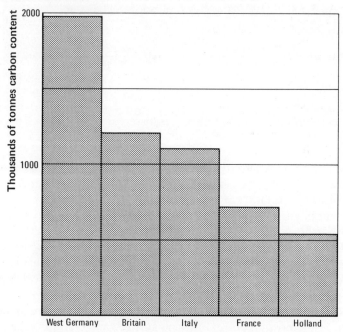

Fig. 3.2   Production of chemicals from petroleum, 1966

Fuels, Explosives and Dyestuffs

The petroleum chemicals industry was founded in the USA soon after the end of World War I and for some 20 years was developed almost exclusively in that country. This was to be expected since the USA has not only always been the largest producer of crude oil in the world, but also the largest producer of motor fuels. The search for ways of producing high grade gasolines for car fuels resulted in the development of cracking processes which in turn led to the production of large quantities of ethylene and propylene.

The first attempt to use these olefins as synthesis gases was made by Carleton Ellis and his colleague Cohen, who founded the Melco Chemical Co. in 1917. In the same year a small plant was put into operation in Bayonne, N.J., for the production of acetone from propylene via iso-propyl alcohol. The process was later taken over by the Standard Oil Co. and a much larger plant incorporating the first successful large-scale fractioning column was installed in 1920. Within three years the annual throughput of this plant had risen to 4 500 hl (100 000 gallons) of iso-propyl alcohol.

(a) propylene $\xrightarrow[\text{(hot)}]{85\% H_2SO_4}$ alkyl sulphate $\xrightarrow[\text{100°C}]{\text{water}}$ iso-propyl alcohol

(b) iso-propyl alcohol $\xrightarrow[\text{250°C}]{\text{metal catalyst}}$ acetone

Almost simultaneously with Carleton Ellis' successful attempts to use propylene as an industrial synthesis gas, another American chemist George O. Curme was working at the Mellon Institute in Pittsburgh in an attempt to produce acetylene from petroleum hydrocarbons. In 1917 Curme developed a process for gasifying middle oil distillates using a powerful submerged electric arc giving a gaseous mixture comprising almost 25% acetylene together with a high proportion of ethylene. However, the production of acetylene by this method proved uneconomic compared with the carbide process and it was dropped. Curme had meanwhile become interested in the possibilities of ethylene as a raw material for organic synthesis by converting it into ethylene chlorohydrin.

$$\underset{\text{ethylene}}{\begin{array}{c}CH_2\\ \|\\ CH_2\end{array}} \xrightarrow{HOCl} \underset{\substack{\text{ethylene}\\\text{chlorohydrin}}}{\begin{array}{c}CH_2-Cl\\|\\CH_2-OH\end{array}}$$

To develop processes of this kind a plant was built at South Charleston (West Virginia) which, by 1925, was producing a range of ethylene derivatives such as ethylene glycol and ethylene dichloride, the latter being used in the 1930's for the production of vinyl chloride and ethylene glycol as an anti-freeze. By the 1930's ethanol was also being produced at the Charleston plant by the sulphuric acid process already in use for the preparation of iso-propyl alcohol from propylene. This opened up a new field for the synthesis of valuable compounds such as acetic acid and acetaldehyde.

Another product which came into prominence in the 1930's was ethylene oxide, the production of which was pioneered in Paris at the laboratories of the Société Française de Catalyse Généralisée using vapour phase catalytic oxidation of ethylene. Antimony catalysts were initially used but these were later replaced by the silver catalysts used today.

A quarter of a century later another metal-catalysed oxidation reaction was developed which involved passing ethylene into an aqueous solution of palladium chloride producing acetaldehyde in good yield. Surprisingly this reaction had been noted by the American chemist F. C. Phillips as early as 1894, but the first large plant using this reaction was not built until 1960. A variation of this process has recently been used by the Russians to produce the valuable monomer vinyl acetate directly from ethylene using solutions of palladium salts in acetic acid.

Another advance in the use of ethylene as a starting material for organic syntheses was the discovery in 1930 that the Fischer Tropsch process could be modified by the addition of ethylene to the reactants. Fischer and Tropsch had shown in 1923 that when synthesis gas (a mixture of hydrogen and carbon monoxide in the ratio of 2:1) was passed over an alkaline iron catalyst at high pressure and temperature, it produced a mixture of straight chain hydrocarbons and certain oxidation products such as aldehydes ('Synthol'). The development of more effective cobalt/thorium oxide catalysts allowed the process to be carried out at lower pressures and temperatures but the 'Synthol' in this case contained no oxidation products of hydrocarbons. The addition of ethylene to the synthesis gas resulted in the reappearance of aldehydes and alcohols in the reaction mixture.

The commercial possibilities of this reaction led to a systematic study of the effect of olefin addition to synthesis gas in Fischer Tropsch reactions by Roelen of Ruhrchemie A.G. in 1937. Experiments carried out in the summer of 1938 included the heating of an equal mixture of ethylene, carbon monoxide and hydrogen at 10 MN/m^2 (100 atmospheres) pressure over a cobalt/thorium catalyst which resulted in the formation of a dark red liquid containing a high proportion of propionaldehyde.

$$3\,\begin{array}{c}CH_2\\ \|\\ CH_2\end{array} + 2CO + 2H_2 \xrightarrow[10\,MN/m^2\,(100\,atm.)]{Co/Th\,catalyst} \begin{array}{c}CH_3\\ |\\ CH_2\\ |\\ CHO\end{array} + \begin{array}{c}C_2H_5\\ \diagdown\\ \diagup\\ C_2H_5\end{array}\!\!C=O$$

ethylene   propionaldehyde   diethyl ketone
                                                                     (70%)   (30%)

As the result of these experiments the Germans developed the 'Oxo' process during World War II. This was designed to produce aldehydes—and hence alcohols by hydrogenation. The strange name 'Oxo' is derived from the German term used to describe compounds containing the carbonyl group ($>$C=O). Mixtures of hydrogen, carbon monoxide and olefins at a temperature of 90–150°C and 40 MN/m^2 (400 atmospheres) pressure were passed over a cobalt catalyst supported upon pumice.

The industrial importance of the oxidation products of hydrocarbons stimulated research into the possibilities of producing such compounds by the direct oxidation of short chain hydrocarbons. The idea was not new, a patent for the production of mixed fatty acids from middle oil distillates by oxidation had been taken out by Schaal of Württemburg in the 1880's. The first commercial development was by Walker of the Cities Service Oil Co. as the result of his attempts to deoxygenate wet natural gas at the Tallant oil plant in Oklahoma in 1926. Small quantities of oxygen entrained in the natural gas had produced corrosion in pipelines, and their removal by passing the gas at pressure over a heated catalyst had unexpectedly produced methanol, formaldehyde and metaldehyde.

Chemicals from Coal and Petroleum

The scope of the process was extended by the Celanese Corporation who built a plant at Pampa in Texas for the production of acetic acid by direct oxidation of n-butane in the liquid phase. The annual production of this plant was rapidly built up to 150 000 tonnes of acetic acid, a large proportion of which was used in the production of cellulose acetate.

$$CH_3CH_2CH_2CH_3 \xrightarrow[6.5 \text{ MN/m}^2 \text{ (65 atm.)} + 200°C]{+ \text{ atmospheric oxygen}} CH_3COOH + H_2O$$
$$\text{n-butane} \qquad\qquad\qquad\qquad\qquad \text{acetic acid}$$

After World War II the Distillers Co. Ltd (DCL) developed a process for the direct oxidation of light petroleum naphthas to acetic acid. Smaller amounts of other acids such as formic, propionic and succinic acid were also produced as by-products.

DCL, also in the immediate post-war period, pioneered the production of acrolein by vapour phase oxidation of propylene using a mixed catalyst of antimony and tin oxides. The belief of DCL that acrolein would prove an important chemical intermediate was soon borne out by the number of syntheses which were worked out once the supply of acrolein became established.

$$\begin{array}{c}CH_3\\|\\CH\\||\\CH_2\end{array} \xrightarrow[\text{Sb/Sn oxide catalyst}]{+ O_2} \begin{array}{c}CHO\\|\\CH\\||\\CH_2\end{array} \begin{array}{c}\xrightarrow{\text{oxidation}} CH_2{=}CH{-}C{\overset{\displaystyle O}{\underset{\displaystyle OH}{\diagdown\!\!\!\!\!\diagup}}}\\ \text{acrylic acid}\\ \xrightarrow{\text{reduction}} CH_2{=}CH{-}CH_2OH\\ \text{allyl alcohol}\end{array}$$
$$\text{propylene} \qquad\qquad \text{acrolein}$$

Important amongst these was the Shell Chemical Co. synthesis of glycerol, acrolein being first reduced to allyl alcohol and then hydroxylated using hydrogen peroxide. Later successful commercial processes were developed by DCL to make acrylonitrile in one stage from propylene, air and ammonia.

$$CH_3CH{=}CH_2 \xrightarrow[\text{catalyst } 450-500°C]{+ NH_3 + \text{atmos. } O_2} CH_2{=}CHCN$$
$$\text{propylene} \qquad\qquad\qquad \text{acrylonitrile}$$

Within the last few years other olefins have been oxidized using the techniques which had been found successful with ethylene and propylene. An interesting example has been the catalytic oxidative dehydrogenation of n-butylenes which does not yield the corresponding aldehyde (crotonaldehyde) or ketone (ethyl vinyl ketone) as might be expected, but butadiene.

$$\begin{array}{c} CH_2{=}CHCH_2CH_3 \\ CH_3CH{=}CHCH_3 \end{array} \xrightarrow[\text{Sb/Sn catalyst}]{+\,O_2} CH_2{=}CHCH{=}CH_2$$

mixture of n-butylenes — butadiene

Attempts have also been made to use catalysts to yield specific oxidation products from petroleum hydrocarbons, using oxygen or oxidizing agents such as peroxides and peracids. Thus boric acid (H_3BO_3) has been used by the Russians to promote the oxidation of paraffin hydrocarbons to alcohols. There also seems to be the likelihood of further advances in the co-oxidation of petroleum hydrocarbons with inorganic gases such as ammonia, sulphur dioxide and hydrogen chloride.

RAW MATERIALS FOR PETROLEUM CHEMICALS

Before describing in detail the manufacture of specific chemicals it is useful to review the raw material sources available. These include natural and refinery gases, liquid hydrocarbons and, to a

Paraffin wax plant at Stanlow refinery (Cheshire). On the left is the high vacuum distillation unit, with the solvent recovery plant on the right

Chemicals from Coal and Petroleum

limited extent, paraffin wax. From these primary products are derived the secondary products which the petroleum chemist uses as his building blocks in synthesizing a wide range of chemicals, including weedkillers, detergents, plastics, cosmetics, fibres, rubbers and paints. Most useful as starting points for synthesis are acetylene and a handful of short chain paraffins (alkanes) and olefins (alkenes) such as methane, ethylene, propylene and butadiene.

Natural gas is an important synthetic raw material in the USA but not in Europe. This is partly due to the less abundant European supplies of gas but mainly because of its composition. Much of the natural gas found in North America is 'wet' (i.e. contains a mixture of gaseous and liquid hydrocarbons). The natural gas from European sources is virtually a mixture of methane and nitrogen with only very small amounts of other substances. Methane can be cracked at high temperature to produce acetylene or steam reformed to yield synthesis gas but does not make a satisfactory feedstock for the production of the more useful olefins.

Refinery gas contains a range of C_1–C_{14} hydrocarbons and hydrogen produced during the distillation, cracking and reforming of petroleum crude. Unless the refinery is a large one it is not usually economical to separate these components, nor is production high enough to justify the building of chemical plant to use them. Most of the valuable olefins in refinery gas are produced during catalytic cracking processes. The short chain paraffins C_1–C_4 are separated from the gasoline fraction of crude oil by distillation—a process termed stabilization. Propane and butane are also produced in significant quantities by combined catalytic cracking and hydrogenation (hydrocracking).

Liquid hydrocarbons are increasingly being used for the production of acetylene and alkenes. The steam cracking of naphtha (low boiling petroleum distillate) is a common route for the manufacture of ethylene, propylene and butadiene. It will be noted that this type of thermal cracking differs from the cracking process used to produce motor fuels in which the desired end-products are liquid.

The steam cracking of wax which is recovered from crude oil by solvent extraction techniques is carried out at a temperature of around 500°C. Slightly elevated pressures are used and the cracking time is usually of the order of a few seconds. A mixture of olefins with chain lengths from C_5–C_{25} is produced together with a small quantity of gas and fuel oil. The mixture is fractionated to achieve partial separation of the components and various cuts used for manufacturing detergents or as a feedstock for the Oxo process.

Acetylene

Although the role of acetylene in petroleum chemistry is now secondary in importance to that of the olefins, production is still expanding rapidly. It was expected that consumption in the USA alone would reach half a million tonnes by 1970.

The traditional route for synthesizing acetylene from calcium carbide is obsolete in the UK but is still a major source of production in countries such as Japan and those having plentiful supplies of hydro-electricity. In Europe production is principally from methane or natural gas. This is subjected momentarily to a temperature of 1 200–1 500°C using an electric arc (Hüls process) or cracking furnace (Wulff process) and the product rapidly quenched.

$$2CH_4 \xrightleftharpoons{+ \text{energy}} CH{\equiv}CH + 3H_2$$

Acetylene is removed from the cracked gas mixture using a selective solvent such as dimethyl formamide or liquid ammonia. In the extensively used Sachsse process, part of the hydrocarbon feedstock is allowed to burn off in oxygen to provide the necessary energy for the acetylene synthesis. Again rapid quenching is necessary to prevent decomposition of the acetylene, and methyl pyrrolidone is used as the extractive solvent. Both the Wulff and Sachsse processes have been modified to use liquid hydrocarbons such as naphtha. Badische Anilin und Soda Fabrik (BASF) have proposed direct quenching of a crude oil flame using an oxygen burner submerged in the fuel.

The value of acetylene in synthesis lies in its highly unsaturated nature. It is very reactive and readily forms addition compounds. The problem is to control its activity and avoid the formation of unwanted side products. Normally special catalysts and operating conditions are employed to give a maximum yield of the required derivative.

Ethylene and Other Olefins

Among the raw materials required for the synthesis of petroleum chemicals ethylene and propylene stand supreme. A measure of their importance is reflected in the US production figures for 1968 which amounted to $5\frac{1}{4}$ million tonnes of ethylene and $3\frac{1}{2}$ million tonnes of propylene. These huge tonnages of olefin gas indicate that their production as by-products of refinery cracking processes is now quite inadequate to satisfy demand. Special cracking operations are therefore carried out currently in the UK with the express purpose of producing olefins. This has the advantage over reliance on refinery gases that the requirements for specific olefins can more easily be met, since the production of cracked gases for the synthesis of petroleum chemicals can be carried out without interfering with the general refinery programme.

The nature of the end-product produced by thermal cracking depends upon the composition of the feedstock and the operating conditions. A high yield of olefin gases is achieved by steam cracking light naphthas at a temperature of 700–900°C and at low pressure. Besides preventing the deposition of carbon (coking) the presence of steam reduces the partial pressure of the oil vapour and raises the olefin content of the cracked product. A steam to hydrocarbon ratio of 1:2 is commonly used.

The hydrocarbon feedstock is first vaporized and then mixed with steam before passing through the cracking furnace in which it is rapidly heated to 750–900°C. After a period of about a second at this temperature the cracked gas is rapidly cooled (quenched) by passing it through an oil-cooled heat exchanger and the heat extracted at this stage is used for steam raising. Further cooling to a temperature of about 130°C is then carried out using a cold oil spray.

The low-boiling fraction is removed by distillation and cooled to remove water and gasoline, the residual oil being recycled. The cracked gas mixture is then compressed and washed with aqueous alkali to remove carbon dioxide and hydrogen sulphide. After drying with anhydrous alumina the gas is then liquefied by reducing the temperature to $-100°C$ at a pressure of 4 MN/m^2 (40 atmospheres).

A view of the ethylene plant at the Baglan Bay, Port Talbot factory of BP Chemicals (UK) Ltd

The liquefied cracked gas is next fractionated using a battery of distillation columns. Four main fractions are taken off:

(a) Hydrogen and methane
(b) Ethane and ethylene
(c) Propane and propylene
(d) A mixture of C_4 hydrocarbons

This leaves a residue of longer chain hydrocarbons which is mixed with the gasoline fraction and used in the manufacture of motor fuel. Further fractionation is carried out to obtain individual alkanes and alkenes from the primary fractions.

	Japan		UK		W. Germany		France		Italy	
	1965	1970	1965	1970	1965	1970	1965	1970	1965	1970
Total amount of ethylene used (thousand tonnes)	800	1 750	495	990	555	1 340	173	620	225	700
Polyethylene	53	54	52	55	36	44	48	49	54	52
Ethylene oxide	10	9	17	14	21	12	27	10	11	5
Ethanol	18	13	15	8	27	17	9	8	—	6
Styrene	7	7	9	8	11	8	9	10	9	13
Ethyl chloride / Ethylene dichloride	7	11	—	—	1	10	1	20	24	22
Others	5	6	7	15	4	9	6	3	2	2

Production and usage of ethylene (Japan and Europe)

(Percentage ethylene consumed)

During 1969 ICI started up the biggest single stream olefin plant in the world on Teesside at a cost of £15 million. The plant is designed to produce 450 000 tonnes of ethylene a year. This is to be used in the manufacture of polyethylene, PVC, detergents and antifreeze. Another giant steam-cracking plant at Baglan Bay (Port Talbot) with an annual output of 340 000 tonnes of ethylene, 200 000 tonnes of propylene and 60 000 tonnes of butadiene is also being erected by BP Chemicals (UK) Ltd and should be on stream by 1970. The naphtha feedstock of this plant will be supplied from the nearby BP refinery at Llandarcy and the gasoline produced will be returned to the refinery.

Ethylene is now produced in enormous quantities on a worldwide scale by steam cracking naphtha. Production in the UK was expected to approach 1 million tonnes by 1970, and in the major industrial countries there is a growth rate of about 12% per year in output.

Its versatility and relative cheapness have resulted in a rapid growth in consumption in Western Europe from 2·3 million tonnes in 1965 to 4·8 million tonnes in 1970. Consumption is expected to double again in the period 1970–5. In order to give greater flexibility to companies using this raw material an ethylene pipeline grid has been proposed for the UK comparable to the proposed German/Dutch grid. Integration has already begun with the construction of a $6\frac{1}{2}$-km, 152-mm (4-mile, 6-in) main to link the ICI Pennine pipeline from Teesside to Runcorn with Shell's Stanlow/Carrington line in Cheshire. The next logical stage would appear to be the extension of this network to include the BP Baglan Bay naphtha cracker and the ICI ethylene oxide plant at Severnside which is already linked to the Esso refinery at Fawley.

THE PRODUCTION OF PETROLEUM CHEMICALS

The growing use of petroleum as a raw material for the production of organic chemicals has necessitated the use of novel techniques and the development of specialized plant. Production of petroleum chemicals on an economic basis has made continuous large-scale processing inevitable. This in turn has made greater demands on the chemical engineer who has had to design installations able to with-

Chemicals from Coal and Petroleum 123

stand high temperatures and pressures and often corrosive chemicals. It has also been necessary to use new catalysts and control techniques in order to provide the desired processing conditions and avoid unwanted side reactions. In addition a decision usually has to be made as to the most economical choice of feedstock. Thus the process can be based on a highly specialized starting material, which is expensive but gives a high yield; or a cheaper crude feed can be used which produces a range of products which have to be separated. These special features of the petroleum chemical industry can best be exemplified by dealing with the manufacture of a specific product in some detail.

The commercial production of acetic acid since the turn of the century provides an excellent illustration of the development of petroleum technology in this way.

Acetic acid was first encountered as a product of the souring of wine, and was certainly known to the ancient Egyptians. Later, during the middle ages it was produced in a crude form as pyroligneous acid during the process of charcoal burning. By neutralizing the pyroligneous acid with lime, calcium acetate was precipitated, from which acetic acid could be recovered by treatment with sulphuric acid. This was a time-consuming and inefficient process, however, and it was not until calcium carbide became available that acetic acid was synthesized on a commercial scale. Acetylene could be produced by treating the calcium carbide with water, and the gas passed into an aqueous solution of mercurous and mercuric sulphates to form acetaldehyde. This could then be readily oxidized to acetic acid.

$$HC\equiv CH + HOH \xrightarrow{\text{mercuric/mercurous sulphate}} \underset{\text{vinyl alcohol}}{H_2C=CHOH} \longrightarrow \underset{\text{acetaldehyde}}{CH_3CHO}$$

The large-scale production of acetic acid dates from World War I, however, when it was required as an intermediary in the production of acetone and cellulose acetate. These were in demand for the manufacture of cordite and aircraft dope. After the war the production of acetic acid waned and it was not until the 1930's that large-scale production was resumed by the Distillers Co. and British Celanese in order to supply the growing demand for cellulose acetate. The earlier acetylene/acetaldehyde route was then aban-

Fuels, Explosives and Dyestuffs

	Before 1880	1880–1930
Stage 1	distillation of wood → charcoal ↓	coke + limestone ↓
Stage 2	pyroligneous acid + milk of lime ↓	calcium carbide + water ↓
Stage 3	calcium acetate + sulphuric acid ↓	acetylene, hydration + water ↓
Stage 4	ACETIC ACID + calcium sulphate ↓ distil to recover acid	acetaldehyde, oxidation + oxygen ↓ ACETIC ACID
Stage 5		

History of acetic acid

doned in favour of catalytic dehydrogenation of ethanol. In this process ethanol vapour and air were passed over a silver-coated cuprous oxide catalyst at 400–450°C.

$$CH_3CH_2OH \xrightarrow[400-450°C]{air + silver\ catalyst} \begin{cases} CH_3CHO\ (45\%)\ \text{acetaldehyde} \\ CH_3COOH\ (45\%)\ \text{acetic acid} \end{cases}$$

ethanol

Chemicals from Coal and Petroleum

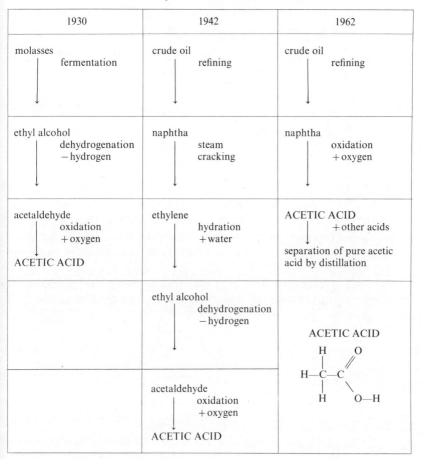

production in the UK

Although the ethanol was at first produced by fermentation, a plant was built in the UK in 1942 by British Celanese to synthesize alcohol from ethylene by hydration. The ethylene feedstock was produced by vacuum cracking hydrocarbon vapour to give a mixed yield of olefins. After separation the ethylene stream was absorbed in concentrated sulphuric acid giving a mixture of monoethyl and diethyl sulphates which was hydrated with the appropriate volume of water. The resulting ethanol was then stripped from the by-products using steam.

$$H_2C=CH_2 + H_2SO_4 \xrightarrow[2\cdot 4\ MN/m^2\ (24\ atm.)]{55-75°C} \begin{cases} C_2H_5OSO_2OH & \text{monoethyl sulphate} \\ \begin{matrix}C_2H_5O\\C_2H_5O\end{matrix}\!\!>\!SO_2 & \text{diethyl sulphate}\end{cases}$$

$$\left.\begin{matrix}C_2H_5OSO_2OH\\ \begin{matrix}C_2H_5O\\C_2H_5O\end{matrix}\!\!>\!SO_2\end{matrix}\right\} \xrightarrow{+\ water} C_2H_5OH + H_2SO_4$$

Smaller amounts of propylene produced as a co-product with ethylene in the cracker were converted to iso-propanol which could be oxidized to acetone. It was then possible to crack the acetone to form ketene which was absorbed in acetic acid to form acetic anhydride.

It is interesting to note that this was a reversal of the older synthetic process whereby acetic acid was used to produce acetone.

$$\begin{matrix}H_3C\\H_3C\end{matrix}\!\!>\!CH(OH) \xrightarrow[500°C\ +\ 0\cdot 3\ MN/m^2\ (3\ atm.)]{\text{catalytic oxidation}} \begin{matrix}H_3C\\H_3C\end{matrix}\!\!>\!C=O \xrightarrow{\text{thermal cracking}} \begin{matrix}H_2C\\ \end{matrix}\!\!>\!C=O$$
iso-propanol   acetone   ketene

$$H_2C=C=O + CH_3COOH \longrightarrow \begin{matrix}CH_3C\\ \\CH_3C\end{matrix}\!\!\begin{matrix}\diagdown O\\ \\ \diagup O\end{matrix}\!\!\!\!\begin{matrix}=O\\ \\=O\end{matrix}$$
acetic acid   acetic anhydride

By 1956 another important development in the history of acetic acid manufacture was the discovery that ethylene could be directly oxidized to acetaldehyde using a palladium catalyst and a copper salt as an oxygen carrier. In the Hoechst process, ethylene and oxygen are passed through the catalyst solution in a reaction tower. Some of the acetaldehyde formed leaves the reactor with unchanged oxygen and ethylene while the rest remains dissolved in the catalyst solution. After recycling the yield is about 95% theoretical. A

Chemicals from Coal and Petroleum

variation on this process (Wacker process) uses a mixture of air and ethylene.

$$CH_2{=}CH_2 \xrightarrow[PdCl_2/CuCl_2]{air\ or\ O_2} CH_3 \cdot CHO \text{ (acetaldehyde)}$$

The success of the Wacker-Hoechst process due to its economy of operation has led to a rapid development of plant capacity since the first installation was commissioned in 1960. World production of acetaldehyde by this route had exceeded half a million tonnes by 1966. In the UK the Distillers Company developed a commercial process for converting acetaldehyde produced in this way directly to acetic anhydride using a copper/cobalt catalyst.

After World War II a considerable amount of research was carried out to explore the possibility of producing acetaldehyde by non-catalytic gas phase oxidation of hydrocarbons. As a result a process was evolved in which mixtures of propane and n-butane, recovered from natural gas or light refinery gas, were reacted with air or oxygen at temperatures of around 450°C and at a pressure of 0·7–1 MN/m^2 (7–10 atmospheres). The reaction yields a complex mixture comprising mainly low molecular weight ketones, aldehydes and alcohols. Acetaldehyde is a major product and is separated by distillation in sufficient purity to permit oxidation to acetic acid.

Attempts have also been made to oxidize paraffin feedstocks in the liquid phase at temperatures between 110–120°C. Again one of the problems encountered is the extremely complex mixture of end-products. This is principally due to the fact that the primary oxidation products are more easily oxidized than the paraffin feedstock. This results in a succession of multiple reactions and a wide variety of products including acids, alcohols, esters, aldehydes, ketones, lactones, water, carbon monoxide and carbon dioxide.

Because of the difficulty of controlling and separating the products yielded by propane/butane oxidation, a liquid phase air oxidation process has been developed based upon light naphtha. As this is one of the primary distillation products of crude petroleum it is much cheaper than n-butane outside the USA. The stream used would normally be a very low octane component of gasoline and would therefore have little commercial value other than as a fuel. The process was worked out after intensive research at what then

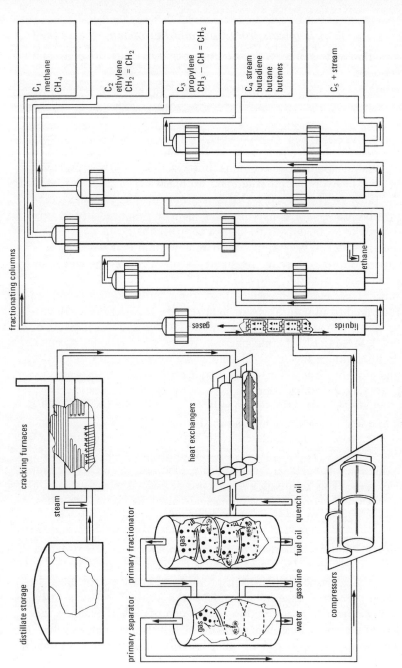

Fig. 3.3 Organic chemicals from petroleum

Chemicals from Coal and Petroleum 129

were the DCL laboratories at Epsom in which gas chromatography proved invaluable in analysing the complex mixture of reaction products obtained. Pilot plant tests carried out at Tonbridge (Kent) were used to help overcome corrosion and reaction control problems

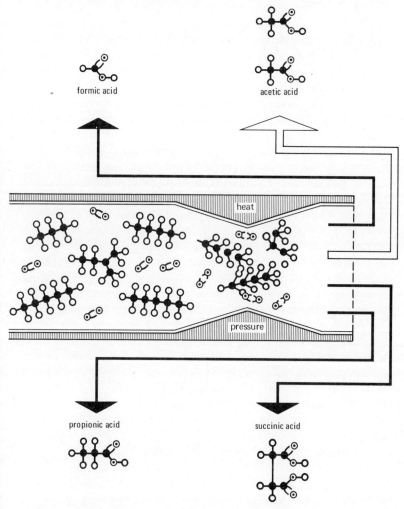

Fig. 3.4 Diagrammatic representation of the DCL acetic acid process (BP Chemicals process)

and the first large-scale plant commenced production at the Salt End factory (Hull) in March 1962.

The light petroleum distillate feedstock is air-oxidized at high temperature and moderate pressure, the reactor being designed to ensure thorough mixing of the gaseous and liquid phases. The reaction is exothermic and proceeds by a free radical mechanism. Alkyl hydroperoxides and ethyl hydroperoxide radicals are formed initially and these break down at high temperatures to give a variety of primary products which are rapidly oxidized to form carboxylic acids and related compounds. The effluent gases from the reactor are first cooled and any unreacted hydrocarbon is recovered and recycled. The liquid phase containing mainly a mixture of formic, acetic and propionic acids is then distilled to effect separation of the products, water being removed azeotropically. Succinic acid is also recovered as a solid by-product from one of the process streams.

The proportion of acids in the oxidizer liquid phase can be varied by altering the operating conditions within the reactor. For this reason special attention was paid during the pilot plant trials to choice of reactor pressure and temperature. Extensive experimentation was carried out assisted by analogue computer studies to achieve the desired products cheaply, safely and in a pure state. Complete control of the entire plant is maintained from a central control room involving the use of some 120 separate monitoring systems. An interesting feature is the use of oxygen meters to record and measure the concentration of oxygen in the reactor. If the oxygen concentration rises above a certain level or there is a fall in reactor temperature then an automatic safety device shuts down the plant.

A second acetic acid plant using the DCL process (now known as the BP Chemicals process) went into production at Hull during 1967, raising the annual potential output from 15 000 to 90 000 tonnes.

The cost of producing acetic acid in this way is about two-thirds of that based upon the ethylene/acetaldehyde route and the product is in a high state of purity. An important point to note is that loss of yield due to complete combustion of the feedstock to carbon dioxide is not of prime importance because of the cheapness of the petroleum feedstock and utilization of the resulting reactor heat

Chemicals from Coal and Petroleum

in raising steam for the distillation unit. Combustion losses of this kind are, however, very serious in processes using expensive feedstocks such as ethylene.

Demand for acetic acid continues to grow and it is used in the manufacture of solvents, plastics, dyestuffs, pharmaceuticals, paints, preservatives, herbicides and adhesives.

PETROLEUM CHEMICALS AND THEIR USES

The diversity and importance of petroleum-based chemicals has already been mentioned. The impact of the petroleum industry is particularly marked in fields such as plastics, fibres, detergents, rubbers and agricultural chemicals. There follows a brief outline of the principal derivatives of the short chain hydrocarbons used in chemical synthesis.

A view of the DF acetic acid complex at BP Chemicals (UK) Ltd's Salt End factory. Here 90 000 tonnes of acetic acid are produced per annum

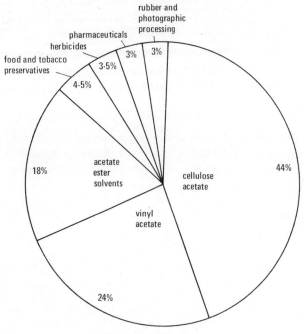

Fig. 3.5   Uses of acetic acid

(a) Acetylene-based Derivatives

Hydrochlorination

Although ethylene is increasingly used to manufacture vinyl chloride, the route via acetylene is still of commercial importance. A mixture of acetylene and dry hydrogen chloride gas is passed over charcoal containing about 10% mercuric chloride at atmospheric pressure and a temperature of 80–100°C.

$$CH\equiv CH \xrightarrow[\text{HgCl}_2/80°C]{+\text{ HCl}} CH_2\!=\!CHCl$$
$$\text{vinyl chloride}$$

Most of the vinyl chloride made is polymerized to the plastic polyvinyl chloride (PVC). Some is chlorinated, however, to form the solvent trichloroethane which can in turn be dehydrochlorinated to form vinylidene chloride. The latter is often copolymerized with PVC ('Saran').

Polypots for polymerizing vinyl chloride

$$CH_2{=}CHCl \xrightarrow{+ Cl_2} CH_2ClCHCl_2 \xrightarrow[\text{lime/80°C}]{- HCl} CH_2{=}CCl_2$$

trichloroethane — vinylidene chloride

Chlorination

Trichloroethylene is prepared by a similar route to trichloroethane. Chlorine is passed through a solution of acetylene in tetrachloroethane using a metal chloride catalyst. The excess tetrachloroethane is catalytically dehydrochlorinated. The resulting trichloroethylene is in demand as a grease solvent and dry cleaning compound.

$$HC{\equiv}CH + 2Cl_2 \xrightarrow{+ FeCl_3\ 80°C} \underset{\underset{CHCl_2}{|}}{CHCl_2} \xrightarrow{+ BaCl_2\ 250°C} \underset{CHCl}{\overset{CCl_2}{\|}}$$

tetrachloroethane — trichloroethylene

Polymerization

Acetylene in the liquid phase forms a dimer (vinyl acetylene) in the presence of a mixed catalyst at 65–70°C. The vinyl acetylene can then be hydrochlorinated to give chloroprene from which neoprene rubber can be produced by polymerization.

$$CH{\equiv}CH + CH{\equiv}CH \xrightarrow{Cu_2Cl_2 + NH_4Cl} CH{\equiv}CCH{=}CH_2 \xrightarrow[Cu_2Cl_2]{+ HCl} CH_2{=}CClCH{=}CH_2$$

vinyl acetylene — chloroprene

Vinyl acetate

Acetic acid reacts with acetylene in the vapour phase at about 200°C using a zinc acetate or mercuric chloride catalyst. After cooling to 0°C the resulting vinyl acetate (VA) is separated and purified by distillation. VA polymerizes readily and is in demand for emulsion paints and adhesives.

$$CH_3COOH + CH\equiv CH \xrightarrow{HgCl_2\ 200°C} CH_3COOCH=CH_2 \longrightarrow PVA$$
$$\text{acetic acid} \qquad\qquad\qquad\qquad \text{vinyl acetate}$$

Acrylonitrile

This involves the addition of hydrogen cyanide to acetylene in the presence of cuprous chloride in hydrochloric acid. Acrylonitrile is widely used in the manufacture of acrylic fibres and nitrile rubber but is more usually synthesized from propylene.

$$CH\equiv CH + HCN \xrightarrow[80°C]{Cu_2Cl_2/HCl} CH_2=CHCN$$
$$\text{acrylonitrile}$$

Acrylic acid and methyl methacrylate

At high pressure (up to 20 MN/m² (200 atmospheres)) carbon monoxide and water react with a solution of acetylene in tetrahydrofuran. Nickel bromide is commonly used as a catalyst. Acrylic esters such as methyl acrylate can be produced by using the appropriate alcohol instead of water.

$$CH\equiv CH + CO + \begin{cases} H_2O \\ CH_3OH \end{cases} \xrightarrow[NiBr_2]{20\ MN/m^2\ (200\ atm.)} \begin{cases} CH_2=CHCOOH \\ \quad \text{acrylic acid} \\ CH_2=CHCOOCH_3 \\ \quad \text{methyl acrylate} \end{cases}$$

(b) Alkane-based Derivatives

1. *Methane*

Synthesis gas. In addition to its use as a source of acetylene, methane is also used in the manufacture of synthesis gas (i.e. a mixture of carbon monoxide and hydrogen). This is used principally for the production of methanol and aldehydes, and as a hydrogen source for the synthesis of ammonia by the Haber process.

Chemicals from Coal and Petroleum

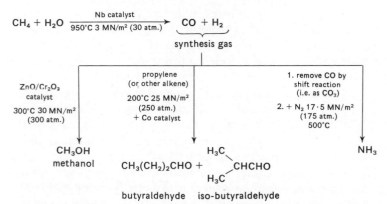

The production of ammonia is of particular interest because it reflects once again the swing away from coal as a base for chemical synthesis. About 80% of the world's ammonia is now derived from petroleum. In addition, although ammonia is one of the few inorganic chemicals made from petroleum, demand is so great that it represents the largest tonnage of any petroleum chemical.

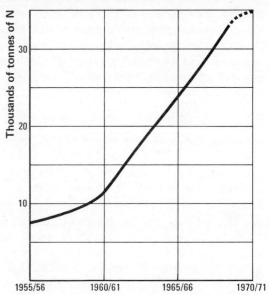

Fig. 3.6 World ammonia production (all sources)

Carbon black. Carbon black, although perhaps less interesting than the other petroleum chemicals, is nevertheless a valuable commodity—its most important use being as a compounding agent in rubber tyres and other toughened forms of rubber, up to 35% by weight being mixed with the rubber before moulding.

Originally carbon black was produced by the channel process, so called because natural gas was burnt at jets which were cooled on sheets of channel iron upon which the carbon collected. Also the thermal process was commonly used, in which a natural gas feedstock was passed into a furnace containing refractory bricks heated to 1 000–1 350°C, cracking producing a coarse granular carbon.

Another type of carbon black (furnace black) is produced by the partial combustion of heavy oil residues at a temperature of 1 200–1 400°C. The hot issuing gases which carry the carbon in suspension are quenched with water and the carbon removed by electro-precipitation. Over half the total carbon black production is now by this method, which was introduced in 1943.

Hydrogen cyanide. The usual source of methane in this reaction is natural gas. This is passed together with ammonia and air over a platinum/rhodium catalyst at 1 000°C (Andrussow process). The gaseous product is cooled, scrubbed to remove excess ammonia and then absorbed in water as hydrocyanic acid. The main outlet for hydrogen cyanide is in the production of acrylonitrile, methyl methacrylate and adiponitrile. An increasing quantity of these materials is now synthesized by alternative routes and demand is slackening.

$$2NH_3 + 3O_2 + 2CH_4 \underset{}{\overset{Pt/Rh\ catalyst\ 1000°C}{\rightleftharpoons}} 2HCN + 6H_2O$$

Chlorination. Despite the difficulty of halogenating paraffins directly, the production of chlorinated methane derivatives using gaseous chlorine is becoming increasingly important. Moderately high temperatures are used and the reaction promoted with ultraviolet light. As methyl chloride is chlorinated more readily than methane, a large excess of the hydrocarbon must be present (10:1) if the former is to be the principal end-product. Chlorinated hydrocarbons are widely used in the production of silicone compounds. The dichloride is used as a paint and grease solvent. Carbon tetra-

Chemicals from Coal and Petroleum

chloride, and to a lesser degree chloroform, are of importance in the production of fluorocarbons.

$$CH_4 \begin{array}{l} \xrightarrow{\text{methane/chlorine 10:1}}_{\text{400°C u/v light}} CH_3Cl \longrightarrow CH_2Cl_2 \\ \text{methyl chloride} \quad \text{methylene dichloride} \\ \xrightarrow{\text{methane/chlorine 1·7:1}} CHCl_3 \longrightarrow CCl_4 \\ \text{chloroform} \quad \text{carbon tetrachloride} \end{array}$$

Other reactions. Methane is used industrially for the production of carbon disulphide by vapour phase reaction with sulphur at 680°C. Small quantities of nitromethane may also be prepared by nitrating methane at 475°C.

2. Ethane and propane

Cracking. Both propane and ethane are cracked to produce ethylene in the USA where they occur in large quantities in association with methane in natural gas. Temperatures of around 800°C are commonly used.

	H_2	CH_4	C_2H_2	C_2H_4	C_2H_6	C_3 hydrocarbons
Cracked ethane	x x x x x	x x	x	x x x x x	x x x x	x
Cracked propane	x x x	x x	x	x x x x	x x	x x x

Comparison of cracked gas compositions from ethane and propane

Key	
>30%	x x x x x
24–30%	x x x x
10–16%	x x x
1–5%	x x
<1%	x

Nitration. Nitroalkanes are produced on a small scale by direct nitration with nitric acid at 400°C and pressures of about 1 MN/m² (10 atmospheres).

$$C_2H_6 \xrightarrow[\text{1 MN/m}^2 \text{ (10 atm.) 400°C}]{+ HNO_3} C_2H_5NO_2 + H_2O$$
$$\text{nitroethane}$$

Halogenation. Chloroethane is produced by vapour phase chlorination of ethane at 300–500°C. The hydrochloric acid produced in this reaction can be used to hydrochlorinate ethylene, thus increasing the yield of chloroethane.

$$\text{C}_2\text{H}_6 \xrightarrow[300-500°\text{C}]{+\text{Cl}_2} \text{C}_2\text{H}_5\text{Cl} + \text{HCl} \xrightarrow{+\text{C}_2\text{H}_4} \text{C}_2\text{H}_5\text{Cl}$$
ethyl chloride

3. Higher alkanes

The DCL process (BP Chemicals process) for oxidizing naphtha distillate in the liquid phase has already been described. This is becoming an increasingly important source of fatty acids. Straight chain alkanes in the C_{10}–C_{15} range are used in the production of alkylbenzenes. These form the basis of one type of biodegradable (i.e. decomposable by bacteria) anionic detergent.

(c) Ethylene-based Derivatives

Polymerization

By far the greater part of the ethylene produced is used in the manufacture of polyethylene. Consumption in the UK for this purpose in 1968 was over 330 000 tonnes, representing almost half the total production (691 000 tonnes). The discovery and early development of polyethylene is described in *High Polymers*, Book 1 in this series. The original process for polymerizing ethylene used very high pressures in the region of 100–250 MN/m^2 (1 000–2 500 atmospheres), the ethylene being maintained at a temperature of 100–300°C and being mixed with 600 ppm of pure oxygen. The resulting viscous liquid polymer was chilled and flaked to give a product of mean density 0·92 g/cm^3. Polyethylene of this type is referred to as 'low density'.

With the development of the Ziegler type catalysts in Italy during the early 1950's, it became possible to polymerize at much lower pressures of 0·6–0·7 MN/m^2 (6–7 atmospheres). This produced a polymer of mean density 0·96 g/cm^3 referred to as 'high density'. A typical Ziegler catalyst contains a titanium tetrachloride and an organometallic reagent such as trimethyl aluminium. The ethylene is dissolved in an inert hydrocarbon solvent such as cyclohexane, which also carries the catalyst. Polymerization is then carried out at a temperature of 100–130°C. The catalyst is deactivated with water or ethanol and the polymer filtered off and converted into granules for ease of handling.

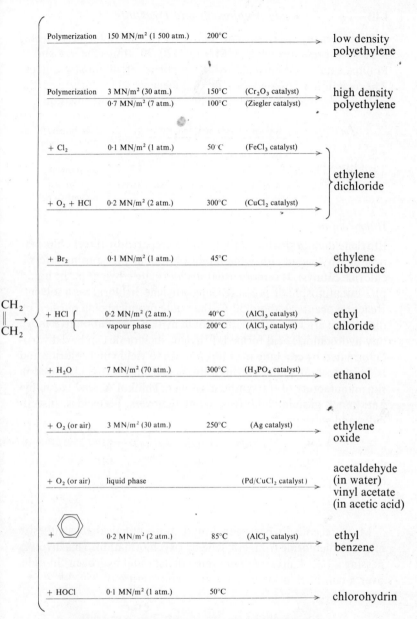

Summary of industrial processes based upon ethylene

High density polyethylene can also be manufactured using moderate pressures of 2–3 MN/m^2 (20–30 atmospheres) and a Phillips catalyst of chromic oxide supported on alumina granules. The reaction is carried out in cyclohexane which dissolves the polymer enabling the catalyst to be removed by centrifugation.

$$CH_2=CH_2 \text{ (ethylene)} \xrightarrow{200\ MN/m^2\ (2000\ atm.)\ 100-300°C\ +\ O_2} \text{low density polyethylene}$$

$$CH_2=CH_2 \xrightarrow{2\ MN/m^2\ (20\ atm.)\ 125-175°C\ \text{Phillips catalyst}} \text{high density polyethylene}$$

$$CH_2=CH_2 \xrightarrow{0.6\ MN/m^2\ (6\ atm.)\ 100-170°C\ \text{Ziegler catalyst}} \text{high density polyethylene}$$

Halogenation

Ethylene dichloride (b.p. 84°C) can be prepared by direct chlorination of ethylene in either the liquid or vapour phase using a ferric chloride catalyst. It is more usual to chlorinate ethylene in the liquid phase at atmospheric pressure using ethylene dichloride as a solvent. Reaction occurs at about 50°C, the crude product being purified by distillation after washing with sodium hydroxide solution to remove any hydrochloric acid formed. Ethylene dichloride can be dehydrochlorinated by cracking over hot pumice to yield vinyl chloride and is an important source of this valuable chemical. It is also used in the manufacture of the synthetic rubber, Thiokol A, lead tetraethyl anti-knock gasoline additives, paint removers, herbicides, insecticides and solvents.

$$\underset{CH_2}{\overset{CH_2}{\|}} + Cl_2 \xrightarrow[50°C]{+\ FeCl_3} \underset{CH_2Cl}{\overset{CH_2Cl}{|}} \xrightarrow[-HCl]{\text{hot pumice}} \underset{CH_2}{\overset{CHCl}{\|}}$$
$$\text{ethylene dichloride} \qquad \text{vinyl chloride}$$

More recently it has been found more practical to synthesize ethylene dichloride from ethylene by oxychlorination. This involves passing a 1:2:4 mixture of oxygen, ethylene and hydrogen chloride, over a cupric chloride catalyst at a temperature of 250–300°C.

$$2\ \underset{CH_2}{\overset{CH_2}{\|}} + 4HCl + O_2 \xrightarrow[250-300°C]{CuCl_2\ \text{catalyst}} 2\ \underset{CH_2Cl}{\overset{CH_2Cl}{|}} + 2H_2O$$

Chemicals from Coal and Petroleum

The method used to prepare ethylene dibromide from ethylene by direct bromination is similar to that used for producing ethylene dichloride. The main outlet for this compound is as an anti-knock additive for gasoline, although it is also used as a fumigant.

Hydrochlorination

Liquid ethylene at high pressure reacts with hydrogen chloride in the presence of an aluminium chloride catalyst at 40°C to give ethyl chloride. The reaction also takes place in the vapour phase but a higher temperature (200°C) is required.

$$\begin{matrix} CH_2 \\ \| \\ CH_2 \end{matrix} + HCl \xrightarrow[\substack{35-40°C \\ + \text{ high} \\ \text{pressure}}]{AlCl_3} \begin{matrix} CH_3 \\ | \\ CH_2Cl \end{matrix} \text{ ethyl chloride}$$

The easily liquefied gas is dispensed from small aerosols as a freezing anaesthetic. It is also used in the preparation of lead tetraethyl.

Hydration

The hydration of ethylene to ethanol is rapidly replacing the traditional fermentation processes and production in the UK is in the region of 125 000 tonnes a year. The earliest process for producing ethanol from ethylene involved absorbing the gas in sulphuric acid at 2·4 MN/m² (24 atmospheres) pressure and then hydrolysing the resulting ethyl sulphates with steam. The alcohol was then stripped from the reaction mixture by steam and purified by distillation.

$$\begin{matrix} CH_2 \\ \| \\ CH_2 \end{matrix} \xrightarrow[55-75°C \ 2\cdot4 \ MN/m^2 \ (24 \ atm.)]{+ H_2SO_4} \left. \begin{matrix} C_2H_5OSO_2OH \\ C_2H_5OSO_2OC_2H_5 \end{matrix} \right\} \xrightarrow{steam} C_2H_5OH + H_2SO_4 \text{ ethanol}$$

mixture of mono- and diethyl sulphates

A catalytic process for directly hydrating ethylene without the use of sulphuric acid was originally used by the Shell Co. at Houston (Texas) in 1947. A pure ethylene feedstock is compressed and mixed with water before passing over a Celite catalyst impregnated with phosphoric acid at 300°C. As the yield per mass is low

142   *Fuels, Explosives and Dyestuffs*

the ethylene is continuously re-cycled over the catalyst. An accumulation of impurities is avoided by 'bleeding off' a certain percentage of the ethylene stream for purification. After scrubbing with sodium hydroxide solution to remove any acids present, aldehydes are converted to the corresponding alcohols by hydrogenation. The alcohols are then stripped off and separated by distillation.

$$\begin{matrix}CH_2\\ \|\\ CH_2\end{matrix} + H_2O \xrightarrow[300°C\ 7\ MN/m^2\ (70\ atm.)]{H_3PO_4/\text{celite catalyst}} \begin{matrix}CH_3\\ |\\ CH_2OH\end{matrix}$$
$$\text{ethanol}$$

Apart from its use as a chemical intermediary, ethanol is a valuable propellant fuel and is widely used as a solvent in the preparation of perfumes, flowers, cosmetics, varnishes, lacquers, inks and as a hydraulic and de-icing fluid. Ethanol can also be oxidized to produce acetaldehyde, although a preferred direct route is by reacting ethylene, water and oxygen in the vapour phase over a palladium/copper catalyst (Wacker process).

Oxidation

Ethylene oxide, one of the most important ethylene derivatives, was discovered over a century ago in Paris by Wurtz. One of the simplest cyclic compounds known, it is a colourless liquid boiling at 12·5°C. It was formerly prepared in two stages. Initially ethylene was passed into a warm solution of chlorine in water contained in an earthenware or rubber-lined steel reactor tower giving rise to a 4–10% solution of ethylene chlorohydrin. The ethylene chlorohydrin was then treated with lime in a horizontal cylinder fitted with baffles and heated with steam, producing a solution which contained about 26% ethylene oxide. Since 1930 direct oxidation of ethylene to ethylene oxide using a silver catalyst has been practised using both oxygen and air, and this process has almost replaced the chlorohydrin route.

(a) $\begin{matrix}CH_2\\ \|\\ CH_2\end{matrix} \xrightarrow[50°C]{+\ Cl_2/H_2O} \begin{matrix}CH_2OH\\ |\\ CH_2Cl\end{matrix} \xrightarrow[100°C]{+\ CaO} \begin{matrix}H_2C\\ |\\ H_2C\end{matrix}\!\!>\!O$

  ethylene   chlorohydrin   ethylene
                                          oxide

(b) $\begin{matrix}CH_2\\ \|\\ CH_2\end{matrix} \xrightarrow[250-300°C\ 3\ MN/m^2\ (30\ atm.)]{+\ O_2/\text{Ag catalyst}} \begin{matrix}H_2C\\ |\\ H_2C\end{matrix}\!\!>\!O$

Chemicals from Coal and Petroleum

As a chemical intermediary it is used to synthesize surface active agents, ethylene glycols, glycol ethers and ethanolamines. Monoethylene glycol is used as an antifreeze and in the manufacture of 'Terylene'. The di- and triethylene glycols are used in the manufacture of alkyd and polyester resins, paints and varnishes. They are also used as humectants and as brake and general hydraulic fluids. The ethanolamines are clear oily liquids which are used in the preparation of cosmetics, polishes, foam stabilizers and acid gas absorbers. Ethylene oxide is also used as a fumigant for cereals, tobacco and certain foodstuffs.

A liquid phase oxidation of ethylene to acetaldehyde in an aqueous medium using a palladium/copper salt catalyst has recently been introduced. If the reaction is carried out in acetic acid, vinyl acetate is formed. Although promising, this route to vinyl acetate is not yet used on a large scale.

Styrene and other alkylation products

The addition of paraffin hydrocarbon chains to chemical structures is an important synthetic process known as alkylation. A number of interesting alkylation reactions occur between olefins and benzene to produce alkyl benzenes by Friedel-Craft type reactions. The liquid phase alkylation of benzene with ethylene to produce ethyl benzene is of outstanding commercial importance. Pure dry benzene is reacted with ethylene containing a little ethyl chloride as a promoter. Solid aluminium chloride is used as a catalyst at a

(*a*) A Friedel-Craft synthesis

benzene + C_2H_5Cl $\xrightarrow{AlCl_3}$ ethyl benzene (C_2H_5) + HCl

(*b*) Direct alkylation with ethylene

benzene ($CH_2=CH_2$) + benzene →
- vapour phase: + H_3PO_4 4 MN/m² (40 atm.) 300°C
- liquid phase: + $AlCl_3$ 0·2 MN/m² (2 atm.) 85–95°C

→ ethyl benzene (C_2H_5)

pressure of 0·2–0·3 MN/m^2 (2–3 atmospheres) and temperature of 85–95°C. Ethyl benzene is separated from the filtered and neutralized reaction mixture by distillation, unchanged benzene being recycled. Vapour phase alkylation of benzene is also carried out using a phosphoric acid catalyst.

The most important commercial outlet for ethylbenzene is in the preparation of styrene which is prepared by direct dehydrogenation over a magnesium or iron oxide catalyst. Superheated steam is used to reduce the partial pressure of the ethylbenzene and maintain reaction temperature at about 600°C. The conversion to styrene is about 40–60% theoretical, unchanged ethylbenzene being separated off and recycled. The separation of styrene by distillation is complicated by the proximity of the boiling points of styrene and ethylbenzene (146°C and 136°C respectively) and the rapid polymerization of styrene at elevated temperatures. Sulphur is usually added as a polymerization inhibitor.

$$\text{ethyl benzene} \xrightarrow[600°C]{\text{MgO catalyst + steam}} \text{styrene}$$

Styrene is used in the production of polystyrene, and copolymers such as the styrene/butadiene rubbers. An inhibitor such as *p*-tert-butyl catechol is often added to prevent spontaneous polymerization of styrene during storage.

A remarkable alkylation reaction occurs if ethylene is passed over aluminium triethyl. The ethylene molecules polymerize to form long chains of from 2–12 carbon atoms which attach themselves to the metal atoms producing higher aluminium tri-alkyl derivatives. Upon oxidation with air the organo-metallic structure breaks down to form aluminium oxide and primary alcohols containing up to 24 carbon atoms, although mostly in the C_6–C_{18} range.

$$\underset{CH_2}{\overset{CH_2}{\|}} \xrightarrow[\text{+ air oxidation}]{Al(C_2H_5)_3} Al_2O_3 + CH_3(CH_2)_nCH_2OH$$

primary alcohol
(where *n* is an integer from 2-22)

(d) Propylene-based Derivatives

After ethylene, propylene is the most important of the 'building bricks' for petroleum chemicals, the annual consumption in the UK in 1968 being 411 000 tonnes. By 1970 this figure was expected to rise to 600 000 tonnes in the UK and $3\frac{1}{2}$ million tonnes in the USA. As might be expected many of the products based upon propylene are analogous to those synthesized from ethylene, such as polypropylene, propylene oxide, iso-propanol and n-butyraldehyde. There are also a number of industrial reactions of propylene which do not have ethylene counterparts.

Polymerization

Although of less importance than polyethylene, polypropylene is made in substantial quantities and production is rapidly growing in both the USA and Europe. Production during 1967 in the UK was of the order of 36 000 tonnes. As in the case of polyethylene, Ziegler type catalysts are used but these are designed to be stereospecific (see *High Polymers*, Book 1 in this series). Aluminium alkyls are used in conjunction with titanium trichloride as catalysts and have the effect of producing a molecular structure with a regular isotactic form. The catalyst is used in hexane at a temperature of 50–100°C and a pressure of 1 MN/m^2 (10 atmospheres). After polymerization has taken place unreacted propylene is allowed to vaporize off by releasing the pressure and the slurry of polypropylene is centrifuged off.

Polypropylene has a higher impact strength and melting point than polyethylene, and is often copolymerized with it. Propylene tetramer-based detergents are no longer used because they are not broken down by bacteria (non-biodegradable).

Halogenation

When propylene is treated with chlorine either substitution or addition can take place. Addition of chlorine across the ethylenic double bond takes place preferentially at lower temperatures to give propylene dichloride. At elevated temperatures however the double bond remains intact and allyl chloride is formed.

$\begin{array}{c} CH_2 \\ \| \\ CH \\ | \\ CH_3 \end{array} \rightarrow$	Reaction	Pressure	Temperature	Catalyst	Product
	Polymerization	0.5 MN/m² (5 atm.)	75°C	(Ziegler catalyst)	polypropylene + copolymers
	Polymerization	1.5–2.5 MN/m² (15–25 atm.)	200°C	(H₃PO₃ catalyst)	tetramer trimer
	+ Cl₂	0.1 MN/m² (1 atm.)	40°C		propylene dichloride
		0.2 MN/m² (2 atm.)	500°C		allyl chloride ↓ glycerol
	+ HOCl	0.1 MN/m² (1 atm.)	35°C		propylene chlorhydrin ↓ propylene oxide
	+ O₂	0.2 MN/m² (2 atm.)	350°C	(CuO catalyst)	acrolein
	+ O₂ + NH₃	0.1 MN/m² (1 atm.)	400°C	(Bi-phosphomolybdate catalyst)	acrylonitrile
	+ H₂SO₄	2–3 MN/m² (20–30 atm.)	25°C	+ hydrolysis	iso-propyl alcohol ↓ acetone ↑
	+ ⌬ (benzene)	2.5 MN/m² (25 atm.)	250°C	(H₃PO₄/AlCl₃ catalyst)	cumene ↓ phenol
	+ CO + H₂	25 MN/m² (250 atm.)	170°C	(Co naphthenate catalyst)	butyraldehydes ↓ butanols
	+ Al(C₂H₅)₃		800°C	(Al silicate)	isoprene
	tetramer + ⌬ (benzene)		85°C	(HF catalyst)	dodecyl benzene

Summary of industrial processes based upon propylene

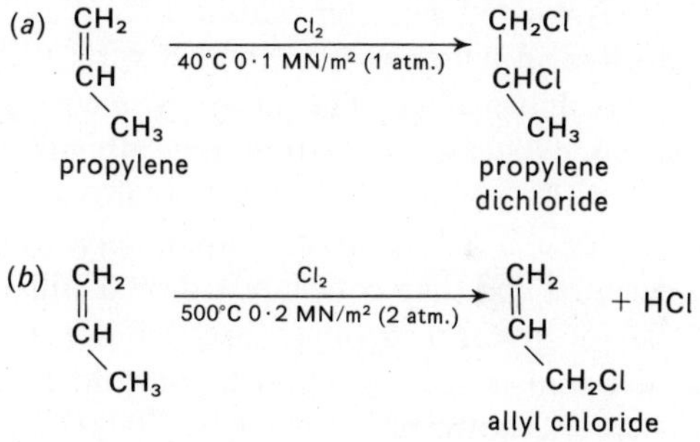

Chemicals from Coal and Petroleum

(a)
$$CH_2{=}CH{-}CH_3 \xrightarrow[40°C\ 0\cdot1\ MN/m^2\ (1\ atm.)]{Cl_2} CH_2Cl{-}CHCl{-}CH_3$$

propylene → propylene dichloride

(b)
$$CH_2{=}CH{-}CH_3 \xrightarrow[500°C\ 0\cdot2\ MN/m^2\ (2\ atm.)]{Cl_2} CH_2{=}CH{-}CH_2Cl + HCl$$

allyl chloride

Until the mid-30's glycerol was produced solely as a by-product of soap manufacture or other fat-splitting processes. Efforts were then made to produce glycerol by a synthetic route using allyl chloride, and a process ('hot chlorination process') was worked out by the Shell Development Co. and first put into operation in the USA in 1948, with a plant having a designed capacity of 25 000 tonnes a year.

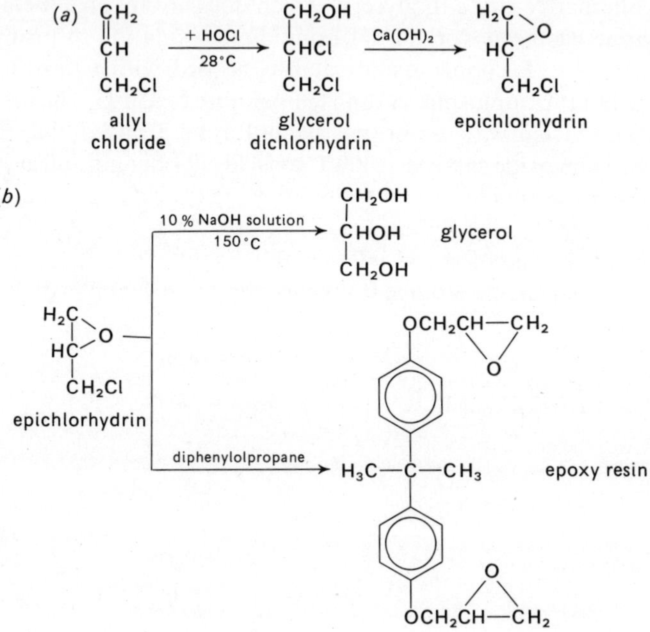

(a)
$$CH_2{=}CH{-}CH_2Cl \xrightarrow[28°C]{+\ HOCl} CH_2OH{-}CHCl{-}CH_2Cl \xrightarrow{Ca(OH)_2} \underset{CH_2Cl}{\underset{|}{HC}}\!\!\overset{H_2C}{\diagdown}\!\!O$$

allyl chloride → glycerol dichlorhydrin → epichlorhydrin

(b) epichlorhydrin

- 10% NaOH solution, 150°C → $CH_2OH{-}CHOH{-}CH_2OH$ glycerol
- diphenylolpropane → epoxy resin

$$OCH_2CH{-}CH_2{\diagdown}O$$ — C$_6$H$_4$ — C(CH$_3$)$_2$ — C$_6$H$_4$ — OCH$_2$CH—CH$_2$ (O)

148 *Fuels, Explosives and Dyestuffs*

Allyl chloride is reacted with hypochlorous acid at 28°C to form glycerol dichlorhydrin which is then treated with a lime slurry to convert it to epichlorhydrin. The latter is recovered from the reaction products by steam distillation and, after fractionating, is hydrolysed with a 10% solution of sodium hydroxide at 150°C in a stirred reactor vessel. The resulting impure glycerol solution is evaporated, desalted and then concentrated by distillation.

The synthesis of glycerol in this country is not at the moment necessary as demand is met by importation and the fat-splitting industries. There is, however, a plant of 20 000 tonnes annual capacity in Holland, and one in France with a capacity of 6 000 tonnes. The American output of synthetic glycerol is currently of the order of 70 000 tonnes, and this amounts to half the annual tonnage used.

Epichlorhydrin is also used in conjunction with diphenylol-propane (from phenol and acetone) to form epoxy resins.

Oxidation

An alternative route to glycerol which does not involve the use of chlorine has been worked out using direct oxidation of propylene to acrolein. A copper oxide catalyst is used with a pressure of 0.2 MN/m^2 (2 atmospheres) and temperature of 350°C. The acrolein is then reacted with iso-propyl alcohol using a mixed zinc oxide/magnesium oxide catalyst at 400°C to yield allyl alcohol and acetone. Hydrogen peroxide is used to hydroxylate the double bond of the allyl alcohol to give glycerol.

Acrolein is a colourless inflammable liquid with a characteristic irritating smell. On storage it polymerizes to a white powder unless

$$\begin{array}{c} CH_2 \\ \parallel \\ CH \\ | \\ CH_3 \end{array} \xrightarrow[350°C \ 0.2 \ MN/m^2 \ (2 \ atm.)]{+ \ O_2 \ (CuO \ catalyst)} \begin{array}{c} CH_2 \\ \parallel \\ CH \\ | \\ CHO \end{array} + \begin{array}{c} H_3C \\ \diagdown \\ CHOH \\ \diagup \\ H_3C \end{array} \xrightarrow[400°C]{(ZnO/MgO)}$$

propylene   acrolein

$$\begin{array}{c} CH_2 \\ \parallel \\ CH \\ | \\ CH_2OH \end{array} \xrightarrow{+ \ H_2O_2} \begin{array}{c} CH_2OH \\ | \\ CHOH \\ | \\ CH_2OH \end{array}$$

allyl alcohol   glycerol

Chemicals from Coal and Petroleum 149

an inhibitor such as hydroquinone is added. Copolymerized with urea and formaldehyde it produces resins which are used in the surface treatment of textiles. It has also been used in the preparation of insecticides and acrylonitrile.

The production of acrylonitrile by the vapour phase interaction of propylene, ammonia and oxygen over a bismuth phosphomolybdate catalyst is carried out at a temperature between 400–500°C and at atmospheric pressure. The process also yields smaller amounts of acetaldehyde, acetic acid and acrylic acids. The relative yields of these by-products can be varied by altering the catalyst and reaction temperature.

$$\begin{array}{c} CH_2 \\ \| \\ CH \\ | \\ CH_3 \end{array} \xrightarrow[400-500°C \; 0 \cdot 1 \; MN/m^2 \; (1 \; atm.)]{+ \; NH_3 + O_2 \; (Bi/phosphomolybdate \; catalyst)} \begin{array}{c} CH_2 \\ \| \\ CH \\ | \\ CN \end{array} + H_2O$$

propylene   acrylontrile

Plant for the continuous manufacture of synthetic glycerol at Pernis (Holland) showing (*left*) part of the reaction section of the allyl chloride unit adjacent to the epichlorhydrin plant

Fuels, Explosives and Dyestuffs

The reactivity of acrylonitrile makes it an important synthetic raw material and the increase in availability of propylene has caused a sharp fall in the price of this derivative which is already in demand for the production of acrylic fibres, thermoplastics and nitrile rubbers. An important property of acrylonitrile is its ability to form addition products with compounds such as alcohols and amines having an active hydrogen atom. Addition products of this type possess a cyanoethyl group ($-CH_2CH_2CN$) which can be readily hydrolysed or hydrogenated.

Propylene oxide is still made almost entirely by the chlorhydrin process although a considerable amount of research has been carried out to find a direct oxidation route. The only commercial process of this type involves the liquid phase reaction of propylene with a hydrocarbon hydroperoxide.

(a) propylene $\xrightarrow{+ HOCl}$ propylene chlorhydrin $\xrightarrow{+ CaO}$ propylene oxide

(b) ethyl benzene $\xrightarrow{\text{air oxidation}}$ ethylbenzene hydroperoxide $\xrightarrow[\text{Mo catalyst}]{+ \text{propylene} \atop 90°C\ 2\ MN/m^2\ (20\ atm.)}$ propylene oxide + phenyl methyl carbinol

Propylene oxide is a useful intermediary for similar compounds to those derived from ethylene oxide, such as propylene glycols and iso-propanolamines. The propylene glycols are sometimes preferred to their ethylene counterparts because of their lower toxicity.

Chemicals from Coal and Petroleum

Hydration

As in the case of ethylene hydration the preparation of iso-propyl alcohol can be carried out indirectly by absorption in sulphuric acid and then hydrolysis of the resulting esters. Unlike ethylene, however, direct hydration is little used because of the tendency of propylene to polymerize under the operating conditions.

Iso-propyl hydrogen sulphate is produced by passing propylene at 2–3 MN/m^2 (20–30 atmospheres) pressure into 80% sulphuric acid at a little above room temperature. The reaction mixture is then treated with steam to promote ester hydrolysis while the temperature is kept below 40°C. The alcohol is then removed by steam stripping and purified by distillation.

$$\begin{array}{c} CH_2 \\ \| \\ CH \\ | \\ CH_3 \end{array} \xrightarrow[2-3 \text{ MN/m}^2 \text{ (20-30 atm.) 25°C}]{+ 80\% \ H_2SO_4} \begin{array}{c} CH_3 \\ | \\ CHOSO_2OH \\ | \\ CH_3 \end{array} \xrightarrow[40°C]{steam} \begin{array}{c} CH_3 \\ | \\ CHOH \\ | \\ CH_3 \end{array}$$

propylene　　　　　　　　　iso-propyl　　　iso-propyl
　　　　　　　　　　　　　hydrogen sulphate　alcohol

The production of iso-propyl alcohol is important because it is an intermediate in the production of acetone. Thus vaporized iso-propyl alcohol at 500°C and 0·3 MN/m^2 (3 atmospheres) is dehydrogenated by passing over a brass, copper or Raney nickel catalyst (150°C).

$$\begin{array}{c} H_3C \\ \diagdown \\ \diagup CHOH \\ H_3C \end{array} \xrightarrow[500°C \ 0·3 \text{ MN/m}^2 \text{ (3 atm.)}]{\text{brass catalyst}} \begin{array}{c} H_3C \\ \diagdown \\ C{=}O \\ \diagup \\ H_3C \end{array}$$

iso-propyl　　　　　　　　　　　acetone
alcohol

Alkylation

Similar methods are used in the alkylation of benzene with propylene as those used with ethylene. Since propylene is more reactive than ethylene, however, operating conditions can be milder and vapour phase processes are convenient.

Cumene is obtained by alkylating benzene with propylene using a phosphoric acid or aluminium chloride catalyst at an operating temperature of 250°C and pressure of 2·5 MN/m^2 (25 atmospheres). During the war cumene prepared in this way was used as an aviation fuel additive but its main use today is as a source of phenol.

Fuels, Explosives and Dyestuffs

The oxidation of cumene to phenol takes place in two stages. An emulsion of cumene and sodium carbonate solution is aerated at 130°C, converting the cumene into the hydroperoxide. The crude hydroperoxide mixture is then acidified with hot dilute sulphuric acid and converted into phenol and acetone.

(a) benzene + CH$_2$=CH–CH$_3$ $\xrightarrow{\text{AlCl}_3,\ 250°C\ 2\cdot5\ \text{MN/m}^2\ (25\ \text{atm.})}$ cumene $\xrightarrow[130°C]{+O_2}$ cumene hydroperoxide

(b) cumene hydroperoxide $\xrightarrow[80–100°C]{\text{dilute H}_2\text{SO}_4}$ phenol + acetone

The production of phenol via cumene has overtaken that from the traditional source—coal tar. Thus by 1966 the annual tonnage of phenol produced from cumene in the UK had reached 78 000 tonnes, representing about 80% of the total annual production from all sources. This is the inevitable result of the rising importance of phenol as a heavy organic chemical and the declining use of coal carbonization processes.

By altering reaction conditions di-iso-propyl benzenes can be prepared by alkylating benzene with propylene. Terephthalic acid (used in the manufacture of 'Terylene') can be prepared by this route.

The lower polymers of propylene such as the trimer and tetramer can also be used to alkylate benzene in the presence of sulphuric acid or hydrogen fluoride. Thus the tetramer (dodecane) forms dodecyl benzene which can be sulphonated to produce an important synthetic detergent base.

(a)

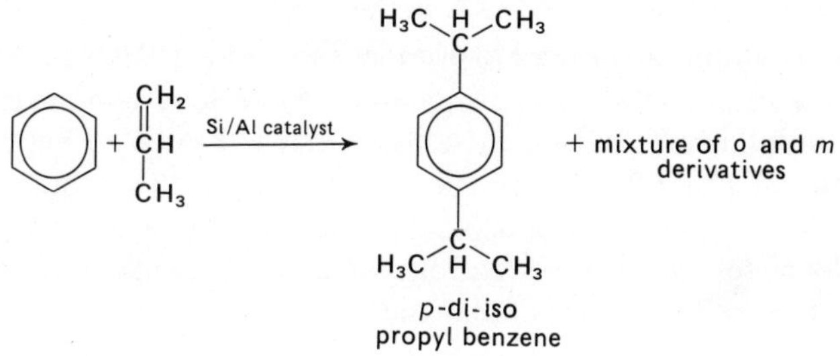

[Reaction: benzene + CH₂=CH(CH₃) —Si/Al catalyst→ p-di-isopropyl benzene + mixture of o and m derivatives]

p-di-iso propyl benzene

(b)

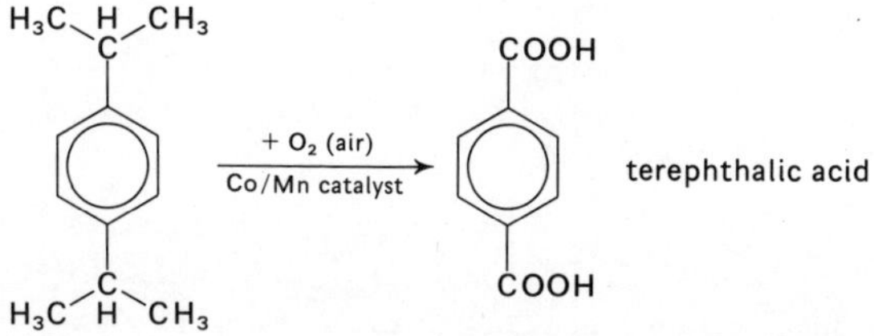

[Reaction: p-di-isopropyl benzene + O₂ (air) —Co/Mn catalyst→ terephthalic acid]

Isoprene

A recent process patented by the Scientific Design Co. is designed to produce isoprene of high purity from propylene. An aluminium tri-alkyl catalyst is used in the first stage to produce 2-methylpent-1-ene which is isomerized to 2-methylpent-2-ene by passage over a second catalyst which is usually aluminium silicate. A mixture of isoprene and methane is produced from this last product by thermal cracking with steam in a stainless steel tube at 700–800°C. This process is only at the pilot plant stage but should provide a valuable addition in the near future to the isoprene prepared by dehydrogenation of C_5 olefins for the synthetic rubber industry.

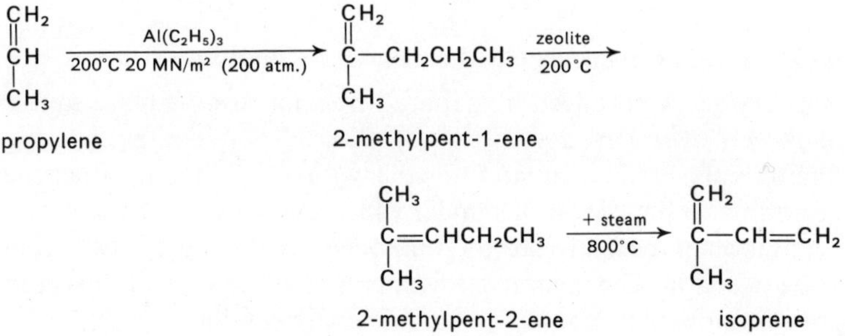

$$\text{propylene} \xrightarrow[200°C\ 20\ MN/m^2\ (200\ atm.)]{Al(C_2H_5)_3} \text{2-methylpent-1-ene} \xrightarrow[200°C]{zeolite}$$

$$\text{2-methylpent-2-ene} \xrightarrow[800°C]{+\ steam} \text{isoprene}$$

Butanols

The Oxo reaction can be used to produce n-butanol and iso-butanol. Propylene is dissolved in a solvent such as acetone containing a cobalt naphthenate catalyst in solution and is heated to between 145° and 175°C at 20 MN/m^2 (200 atmospheres) with synthesis gas (1:1 ratio of carbon monoxide and hydrogen). The resulting butyraldehydes can then be hydrogenated over a copper chromite catalyst to give the corresponding alcohols.

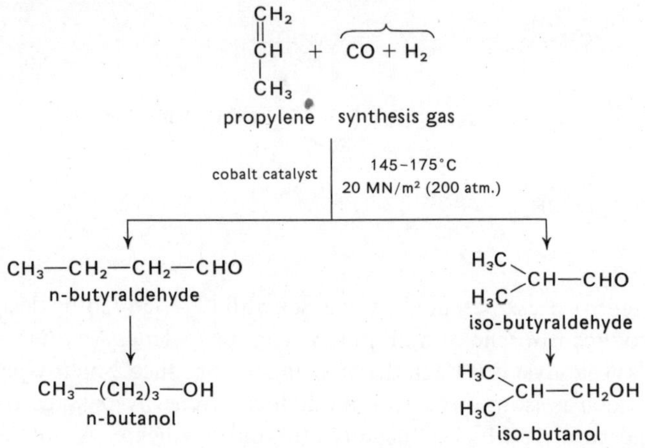

$$\begin{array}{c} CH_2 \\ \parallel \\ CH \\ | \\ CH_3 \end{array} + \overbrace{CO + H_2}$$

propylene   synthesis gas

cobalt catalyst | 145–175°C
20 MN/m^2 (200 atm.)

$CH_3-CH_2-CH_2-CHO$
n-butyraldehyde

↓

$CH_3-(CH_2)_3-OH$
n-butanol

$\begin{array}{c} H_3C \\ \diagdown \\ CH-CHO \\ \diagup \\ H_3C \end{array}$
iso-butyraldehyde

↓

$\begin{array}{c} H_3C \\ \diagdown \\ CH-CH_2OH \\ \diagup \\ H_3C \end{array}$
iso-butanol

Higher alcohols have been produced by this route using propylene polymers such as the trimer and tetramer (i.e. polymers containing 3 and 4 monomer units).

(e) Petroleum Chemicals based on Butylenes and Butadiene

Refinery gases produced by naphtha cracking are the main source of the C_4 hydrocarbons used in synthesis. These contain varying amounts of n- and iso-butane, together with C_4 olefins and butadiene according to the conditions under which cracking was carried out. Separation of the hydrocarbon components is effected by distillation and extraction. The theoretical syntheses based on C_4 hydrocarbons are enormous, but not all are commercially possible.

n-butylenes

The main outlet for the two isomeric n-butylenes is conversion into butadiene by dehydrogenation. Also, hydrolysis with 70% sulphuric acid gives butan-2-ol, which can readily be dehydrogenated, using a zinc or copper catalyst at 350°C, to methyl ethyl ketone (MEK). Both MEK and butan-2-ol are useful solvents and are used in the manufacture of lacquers, and for extraction and crystallization.

$$CH_3-CH_2-CH=CH_2 \quad \text{but-1-ene}$$

$$CH_3-CH=CH-CH_3 \quad \text{but-2-ene}$$

$$\xrightarrow[+ I_2]{\text{catalytic dehydrogenation}} CH_2=CH-CH=CH_2 \quad \text{butadiene}$$

$$\xrightarrow[70\% H_2SO_4]{+ H_2O} CH_3-\underset{\underset{\text{butan-2-ol}}{|}}{\overset{OH}{C}H}-CH_2-CH_3 \xrightarrow[Zn/Cu \; 350°C]{-H_2}$$

$$\underset{H_5C_2}{\overset{H_3C}{>}}C=O \quad \text{MEK}$$

Iso-butylene readily dissolves in 60% sulphuric acid at room temperature and is hydrolysed to tert-butanol which is a useful solvent. At higher temperatures polymerization occurs to give di-iso-butylene together with smaller amounts of the dimer and trimer. These are used as additives and in the manufacture of phenolic resins. Poly-iso-butylenes of high molecular weight (3 000–200 000) are prepared at low temperature (>0°C) using Friedel-Craft type

$$\underset{\underset{\text{iso-butylene}}{}}{\overset{CH_3}{\underset{CH_3}{|}}C=CH_2}$$

$$\xrightarrow{60\% H_2SO_4 \; 20°C} CH_3-\underset{\underset{CH_3}{|}}{\overset{\overset{CH_3}{|}}{C}}-OH \quad \text{tert-butanol}$$

$$\xrightarrow{60\% H_2SO_4 \; 100°C} \text{mixture of di- and tri- iso-butylenes}$$

$$\xrightarrow{50°C + AlCl_3} \text{poly-iso-butylene}$$

catalysts and are used in surface coatings, plasticizers and as viscosity index improvers for lubricating oil ('Vistanex', 'Oppanol'). Butyl rubber is a copolymer of iso-butylenes with 2–5% of Isoprene.

Butadiene

The main rubber producing centres of the Far East were captured by the Japanese in the early 1940's. This triggered off a massive programme of research in the USA into the possibility of synthesizing rubbers on a commercial scale. As a result there was a large increase in the production of butadiene. After the war attempts were made to find outlets for the butadiene capacity which was surplus to peacetime requirements. Although as a result butadiene has been used as an intermediary in a number of synthetic processes, it is still largely used in the production of synthetic rubbers.

Styrene-butadiene rubber (SBR) is produced in the UK by the International Synthetic Rubber Co. (Hythe and Grangemouth), annual output being around the region of 120 000 tonnes. Butadiene is copolymerized with one third of its weight of styrene, forming an aqueous emulsion which is coagulated and processed (see section on 'Elastomers' in *High Polymers*, Book 1 in this series). Over half the butadiene produced is used for this purpose, the main outlet being the manufacture of tyres and footwear. Butadiene is also copolymerized with acrylonitrile to form nitrile rubber (NBR). The direct polymerization of butadiene to polybutadiene (BR) is increasing in importance, alkyl lithium being used as a catalyst. Recently Ziegler type catalysts have been used to raise the '*cis*' type of structures present in the polymer from about 35% to over 95%. This gives a product with properties closely resembling those of natural rubber.

cis 1,4-structure of polybutadiene

trans 1,4-structure of polybutadiene

Butadiene is also used in the synthesis of hexamethylene diamine which is copolymerized with adipic acid in the production of nylon 66.

(a)

$$\underset{\text{butadiene}}{\begin{array}{c}CH_2\\ \|\\ CH\\ |\\ CH\\ \|\\ CH_2\end{array}} \xrightarrow[67-75°C]{+Cl_2} \underset{\text{1,4-dichloro-2-butylene}}{\begin{array}{c}CH_2Cl\\ |\\ CH\\ \|\\ CH\\ |\\ CH_2Cl\end{array}} \xrightarrow[\text{(CuCl}_2\text{ catalyst)}]{+NaCN} \underset{\text{1,4-dicyano-2-butylene}}{\begin{array}{c}CH_2CN\\ |\\ CH\\ \|\\ CH\\ |\\ CH_2Cl\end{array}}$$

(b)

$$\underset{\substack{\text{1,4-dicyano-}\\ \text{2-butylene}}}{\begin{array}{c}CH_2CN\\ |\\ CH\\ \|\\ CH\\ |\\ CH_2CN\end{array}} \xrightarrow[300°C]{+H_2 \text{ (Pd catalyst)}} \underset{\text{adiponitrile}}{\begin{array}{c}CH_2CN\\ |\\ CH_2\\ |\\ CH_2\\ |\\ CH_2CN\end{array}} \xrightarrow[135°C\ 67\ MN/m^2\ (670\ atm.)]{+H_2 \text{ (Co catalyst)}} \underset{\substack{\text{hexamethylene}\\ \text{diamine}}}{\begin{array}{c}CH_2-CH_2NH_2\\ |\\ CH_2\\ |\\ CH_2\\ |\\ CH_2-CH_2NH_2\end{array}}$$

The selective solvent sulfolane which is used in the extraction of aromatics from petroleum distillates is prepared by treating butadiene with sulphur dioxide.

$$\underset{\text{butadiene}}{\begin{array}{c}CH_2\\ \|\\ CH\\ |\\ CH\\ |\\ CH_2\end{array}} \xrightarrow[\text{elevated temperature/pressure}]{+SO_2} \underset{\substack{\text{tetramethylene}\\ \text{sulphone}}}{\begin{array}{c}HC=\!=\!CH\\ H_2C\diagdown\ \diagup CH_2\\ S\\ O_2\end{array}} \xrightarrow{+H_2} \underset{\text{sulfolane}}{\begin{array}{c}\square\\ S\\ O_2\end{array}}$$

(f) Aromatic and other Cyclic Petroleum Chemicals

Traditionally the aromatic hydrocarbons, which have always rated highly as raw materials in the manufacture of organic heavy chemicals, have been obtained by the distillation of coal tar. Within the last few years however, chemists have increasingly turned to oil as the provider of major aromatic compounds, such as benzene, toluene, styrene, xylenes and phenol.

Another ring compound of importance is cyclo-hexane which is separated from petroleum feedstocks by fractional distillation. A major use of this compound is in the preparation of adipic acid and caprolactam—both of which are used in the production of nylon. Cyclo-hexanone is produced as an intermediate product in both cases.

158 Fuels, Explosives and Dyestuffs

cyclohexane → (oxidation (air) 125–160°C, cobalt naphthenate 0·3–1·8 MN/m² (3–18 atm.)) → cyclohexanone

cyclohexanone + NH_2OH at 20°C → cyclo-hexanone oxime → (H_2SO_4) → caprolactam

cyclohexanone + HNO_3, Cu/Vd catalyst, 80°C 0·4 MN/m² (4 atm.) → adipic acid: HOOC–$(CH_2)_4$–COOH

Crudes from certain areas contain naphthenic acids which are carboxylic acids derived from cyclo-paraffins. The metallic salts of these acids are of commercial importance. Thus copper and zinc naphthenates are used as paint 'driers' and fungicides, and calcium naphthenate is a lubricating oil additive.

The isomeric xylenes and toluene are produced by the dehydrogenation of naphthenes during platforming. These are separated from the reformate by selective solvents such as sulfolane or aqueous diethylene glycol.

sulfolane

diethylene glycol: $HOCH_2$–CH_2–O–CH_2–CH_2OH

Toluene is used to manufacture benzene by hydrodealkylation or is oxidized to benzoic acid. If the potassium salt of benzoic acid is heated to 410°C under a pressure of 1 MN/m² (10 atmospheres) of carbon dioxide and with cadmium benzoate as a catalyst, disproportionation takes place. Benzene and potassium terephthalate are formed—the latter is a useful intermediate for the preparation of terephthalic acid ('Terylene' manufacture).

Chemicals from Coal and Petroleum

[Reaction scheme: toluene undergoes hydrodealkylation to benzene; toluene + air + Co catalyst at 120°C, 0·3 MN/m² (3 atm.) gives benzoic acid; benzoic acid + KOH → potassium benzoate; potassium benzoate with Cd benzoate at 410°C, 1 MN/m² (10 atm.), CO₂ gives benzene + potassium terephthalate.]

Terephthalic acid is more usually prepared from *p*-xylene by liquid phase oxidation. Air is used as the oxidizing agent and the reaction proceeds at 200°C and 1·5–3 MN/m² (15–30 atmospheres) pressure in the presence of a cobalt catalyst. *o*-xylene is used as a raw material in the manufacture of phthalic anhydride.

(a) *p*-xylene + O_2 (air), Co catalyst, 200°C 1·5–3 MN/m² (15–30 atm.) → terephthalic acid

(b) *o*-xylene + O_2 (air), V_2O_5 catalyst, 550°C → phthalic anhydride

Product	Tonnes per annum
Acetone	210 000
Acrylonitrile	40 000
Benzene	900 000
Butadiene	190 000
Butyl rubber	24 000
Ethanol (synthetic)	124 000
Ethylene dichloride	540 000
Ethylene oxide	90 000
Ethylene	1 100 000
Iso-propanol	172 000
Methanol	320 000
Nitrile rubber	18 000
Phenol (synthetic)	98 000
Phthalic anhydride	120 000
Polybutadiene	50 000
Polychloroprene	30 000
Polyethylene, high density	74 000
low density	370 000
Polypropylene	75 000
Polystyrene	138 000
Polyvinyl chloride	331 000
Propylene oxide	90 000
Propylene	800 000
Styrene-butadiene rubber	163 000
Styrene	240 000
Vinyl acetate	72 000

United Kingdom plant capacity as at end 1968

Chapter 4
Explosives

Until the upsurgence of organic chemistry around the middle of the last century the only explosive material in use was gunpowder ('blackpowder'). The invention of gunpowder has been at various times credited to the Hindus, Arabs and to the Chinese. The earliest reference to a mixture of saltpetre, sulphur and charcoal, however, appears in a chemical report published in 919 by the philosophical Chinese sect known as Taoists. The inflammability of the mixture was stressed rather than its explosive nature, and it was as a slow-burning propellant that the mixture was first used in 1044—as a filling for bamboo rockets carrying arrows and spears. Later, during the twelfth century, in the Sung Dynasty, the Chinese are again reported to have used gunpowder-filled bamboos as incendiary weapons against the Chin Tartars.

Investigation of the explosive nature of gunpowder is generally credited to the friar, Roger Bacon, during the thirteenth century. Alarmed at his discovery, however, Bacon concealed his formula in a Latin code and a century passed before its resurrection and use in warfare as a propellant. Later, gunpowder mines were used by the British during the siege of Harfleur in 1415.

A number of interesting explosive devices are described by Baptista Porta in his *Twelfth Book of Magick* published in 1558 which he introduces in the following way:

'I shall treat of that dangerous Fire that works wonderful things, which the Vulgar call Artificial Fire . . . there is nothing in the world that more frights or terrifies the minds of men. You have here the compositions of terrible Gunpowder that makes a noise, of Pipes that vomit forth deadly fires—Of fire balls that flie with glittering Fire, and terrifie Troops of Horse-men, and overthrow them.'

By the sixteenth century a number of gunpowder factories had become established in Britain, including the Royal Gunpowder Factory at Waltham Abbey, just north of London. This spot was probably chosen because of its proximity to the charcoal produced in Epping Forest and the water power provided by the River Lea. Factories also sprang up near other charcoal producing areas in Kent and Westmorland.

Early in the seventeenth century there appears the first recorded peaceful use of gunpowder as a blasting agent in the Royal Mines at Schemnitz in Hungary, and later in the Cornish tin mines in 1689. Detonation of the blasting charge at this time was carried out by means of a straw filled with powder, and it was not until 1831 that William Bickford, a Cornish leather merchant, invented the safety fuse. This replaced the straw with a woven cotton tube which could be waterproofed with gutta percha to enable it to burn under water.

Blackpowder is still manufactured on a limited scale for pyrotechnics and safety fuse. A dampened mixture of potassium nitrate, sulphur and wood charcoal is finely ground for several hours and then compressed to form 'press cake'. This is then milled to form grains of the required size, the latter being finally polished in revolving drums. Tests are carried out on the finished product to check its burning speed both in the open and under confinement. The rate of burning is affected by varying the proportions of charcoal and potassium nitrate. When confined, as in safety fuses, the rate of burning is also considerably affected by pressure.

The train of chemical reactions occurring during the combustion of blackpowder is complex and still not fully understood. The oxidation processes primarily responsible for the explosion of ignited blackpowder may be represented by the following composite equation.

$$10KNO_3 + 8C + 3S = 2K_2CO_3 + 3K_2SO_4 + 6CO_2 \uparrow + 5N_2 \uparrow$$

Recent evidence suggests that a secondary reaction also occurs at a much slower rate which involves partial reduction of the residue of potassium salts to potassium sulphide with the evolution of carbon dioxide gas.

Modern safety fuse is manufactured by feeding fine grain gunpowder (blackpowder) into a metal die where it is enclosed by a jute

Explosives

mesh and compressed. After passing through a waterproofing compound such as bitumen the fuse is given an outer covering of varnished cotton. The use of plastic sheathing for safety fuses has the disadvantage that the combustion gases do not vent so evenly and this tends to produce erratic burning. Igniter cord is made by treating a core of textile threads with a mixture of blackpowder dispersed in nitrocellulose dope. This core is then covered with an incendiary layer of nitrocellulose dope to which has been added an oxidizing agent such as potassium nitrate. Finally the cord is sheathed in polythene. Igniter cord is waterproof and will continue to burn even if the outer sheath is pierced. It is often used to ignite several lengths of safety fuse at once and thus give the shotfirer more time to take cover than if the fuses were ignited individually.

Other types of fuse are sometimes used for special purposes, such as the instantaneous fuses used for pyrotechnic 'set pieces' and detonating fuses which are capable of initiating explosions without the use of detonators. An example of the latter (Cordtex) is commonly used in the UK for quarrying.

HIGH EXPLOSIVES

The history of high explosives dates from the discovery of guncotton by Schonbein in 1838. This stimulated interest in the results of nitrating other substances using nitric acid mixtures and led to the discovery of glyceryl trinitrate (nitroglycerine) by Ascanio Sobrero in 1845 at Turin University. Two hundred grammes of Sobrero's first sample are still preserved at the Montecatini plant at Avigliana near Turin, under their original name of 'pyroglycerine'.

$$\begin{array}{c} CH_2OH \\ | \\ CHOH \\ | \\ CH_2OH \end{array} + 3HNO_3 \xrightarrow[20°C]{+H_2SO_4} \begin{array}{c} CH_2ONO_2 \\ | \\ CHONO_2 \\ | \\ CH_2ONO_2 \end{array} + 3H_2O$$

glycerol → glyceryl trinitrate ('nitroglycerine')

The first attempts to utilize glyceryl trinitrate (GTN) as a commercial explosive were carried out by the Nobel family in Sweden and the first patent application was filed by Alfred Nobel on October 14th, 1863. Emmanuel Nobel, the father of Alfred, had been interested in the design of sea mines using gunpowder, and the possibility of using GTN occurred to his son, who subsequently carried out experimental work on these lines in his laboratory in Stockholm. Early setbacks included a disastrous explosion on September 4th, 1865, which killed Alfred's brother and so shocked his father that he never recovered. Another difficulty was the detonation of the new 'blasting oil', as it became known. Unlike the primary explosives and propellants which could be exploded by ignition, the new high explosive compounds had to be detonated by shock, and in 1865 Alfred Nobel patented a metal-cased 'detonating cap' filled with mercury fulminate ($Hg(OCN)_2$).

Meanwhile despite the anxiety of governments and engineers over the dangerous nature of blasting oil, production had started in other countries, notably in Krummel (Germany) in 1865, Massachusetts and San Francisco (USA) in 1867–70, and Ardeer (Scotland) in 1871. Production was often carried out in an unbelievably crude manner. At the Massachusetts factory which had been built to provide blasting oil for the construction of the great Hoosac tunnel, George Mowbray, the owner, describes the nitration of glycerol contained in glass jars into which the nitrating mixture of nitric and sulphuric acids was dripped from stone pitchers over a period of up to two hours.

'The three men, whose duty it is to attend to the mixing process, have each a row of pitchers to watch, walking the whole time up and down, beside them, with thermometer in hand, and, as the nitrous fumes rise from the forming nitroglycerine, they stir the mixture, with a glass tube, in any pitcher that may be giving out too violent fumes.'

Mixing in Mowbray's factory was carried out by the use of compressed air, but in the factory of the Giant Powder Co. in San Francisco mechanical stirring was adopted. Mr Amark, the superintendent, describes how when nitrous fumes developed the two Chinese labourers who operated the mechanical mixer took fright and ran away leaving Amark to work the mixer himself.

In Ardeer the mixing was entrusted to 'hillmen'—so called because their work-sheds were situated on small hills to enable the

GTN to be easily piped away. These hillmen had to have courage and concentration since overheating of the explosive during the exothermic nitration process could easily have resulted in an explosion. In order to prevent them falling asleep in the heavy, sweet, fume-laden atmosphere they were required to sit upon a one-legged stool to ensure a rapid awakening if they dozed off!

At the Swedish factory at Gyttorp on the shores of Lake Vikern, the washed GTN was poured into buckets and taken down to the lake where the washing water was decanted off. The small amounts of explosive which entered the lake apparently caused no anxiety.

Nobel's blasting oil was still plagued by the hazard of accidental explosion and Alfred Nobel appreciated that it was necessary to find a suitable absorbent material for GTN so that it could be handled with complete safety. After a number of failures he found that a diatomaceous earth ('Kieselguhr') would absorb several times its own weight of the oil to form a crumbly plastic mass which could be safely packed and transported. The new material which he termed 'dynamite' (Gk. *dynamos*—power) was patented in 1867 and rapidly grew in popularity.

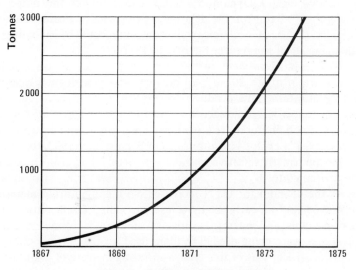

Fig. 4.1   Production of dynamite

Alfred Nobel, writing to *The Times* in February 1868 regarding an accidental explosion of nitroglycerine at Newcastle, complained that accidents such as this resulting from gross carelessness had led to unreasonable criticism of the introduction of this valuable explosive. 'It is high time that the public should know that nitroglycerine has won its battle over prejudice.' He went on to comment that if it was desired to produce a world completely free from danger it was necessary to prohibit the use of steam, fire, poisonous substances, cutting tools, firearms, etc.

In 1875 Nobel produced another explosive by mixing GTN with cellulose nitrate (guncotton) to form a rubber-like jelly which he termed 'blasting gelatine' or 'rubber dynamite'. By varying the amount of cellulose nitrate present, explosive jellies of various consistencies could be obtained and this explosive proved an ideal blasting agent for such operations as quarrying and tunnelling. Even today the gelatines are the most powerful non-nuclear explosives known, having about $1\frac{1}{2}$ times the power of TNT, and for this reason are used as standards to compare the effect of other explosives. By adding materials such as ammonium nitrate and wood flour to a runny gelatine, Nobel also produced an explosive which he termed 'Extra Dynamite', better known under the name of 'Gelignite'. Mixing of the gelatine was originally carried out by hand but the use of mechanical mixers was pioneered by McRoberts, the manager of the Ardeer factory, in 1877.

GTN explosives were further improved in the 1920's by the admixture of about 20–30% of ethylene glycol with the glycerine before nitration. This produced an explosive which remained unfrozen even after prolonged exposure to temperatures as low as $-10°C$. Such mixtures are known as 'low freezing' and are the only types of GTN explosives authorized in Britain since 1932, the latter being abnormally sensitive to shock and highly dangerous to use in the frozen state.

As the demand for GTN increased, so manufacturing methods improved and as early as 1870 lead nitrating vessels, fitted with cooling coils and having a capacity of about a tonne of nitroglycerine, were being used. The plant used for batch production of GTN at Ardeer was of this type, a nitrating mixture of about 45% nitric acid and 55% sulphuric acid being used, and the temperature being kept in the region of 15–20°C by passing chilled calcium chloride solution at −15°C through the cooling coils.

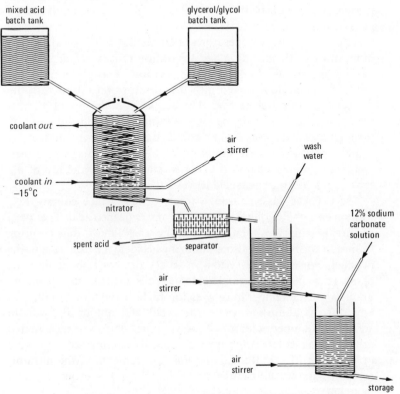

Fig. 4.2   Flowchart showing batch production of nitroglycerine

The first continuous nitration process was patented by an Austrian chemist, Arnold Schmid, in 1927. This not only had the usual advantages of a continuous process over batch production, such as cheapness, compactness, shorter processing time and easier control, but was safer to operate as smaller amounts of GTN were processed at a time. Although initially the continuous process was slow to be adopted, a substantial proportion of present output is by this method, the first Schmid plant being installed at Ardeer in 1937 with a throughput of over a tonne an hour.

The most modern type of plant for continuous production of GTN was designed by Mario Biazzi, a Swiss, in 1935. This shows several improvements on the older Schmid process and is fabricated throughout in polished stainless steel to prevent accumulation of pockets of explosive. The system is compact and can be operated by remote control by an operator seated behind a blast wall, visual monitoring of the nitration and separation processes being carried out by means of closed circuit television. The glycerine/glycol mixture is continuously fed into the nitrator and mixed with the correct ratio of nitrating acid. The temperature during nitration is maintained at 10–15°C by passing sodium nitrate solution at $-5°C$ through a sealed spiral cooling coil, agitation being maintained by a high speed stirrer. An emulsion of GTN and excess acid then enters a spinning separator. GTN is removed from the top of the separator while the spent acid leaves at the bottom.

The GTN is passed through three washers which contain 12% sodium carbonate solution to remove any entrained acid. The speed of stirring emulsifies the GTN with the wash liquor, thus ensuring thorough washing and also minimizing accidental detonation of the product. An elaborate system of safety devices is built into the plant. These include warning lights and bells to attract the operator's attention if the temperature rises too high, if the flow of liquor at any stage is abnormal, or in the event of plant breakdown. In addition a number of interlocks prevent the controls being actuated in the wrong order. Alternative stirring by compressed air occurs in the event of a power failure, and the contents of the nitrator, separator and washing tanks can be dumped into a 'drowning' tank in an emergency.

The final processing and mixing of the GTN takes place in small isolated wooden buildings separated by blast walls. The operatives wear rubber overshoes and the floors are lead-lined to prevent sparks. GTN is conveyed in the plant either along polythene tubing or by means of special plastic tanks mounted on rubber-wheeled trolleys. Mixing is carried out in plastic containers using mechanical mixers or wooden paddles. Gelatinization of guncotton and GTN mixtures occurs after standing for some hours.

Kieselguhr is rarely used in the formulation of blasting gelatines and dynamites in modern practice, other absorbents such as wheat

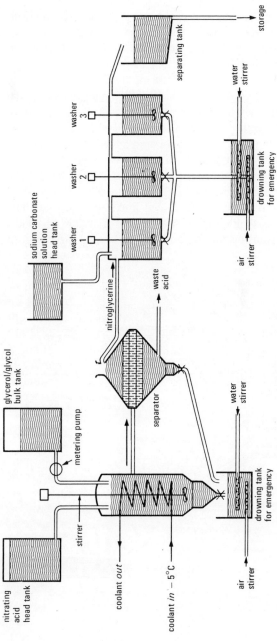

Fig. 4.3 Flowchart showing Biazzi type continuous process for manufacture of nitroglycerine

flour, starch, woodmeal and china clay being preferred. Ammonium nitrate is also often used to prepare 'ammonia dynamite' and 'semi-gelatines'. The inclusion of these materials cheapens the product and produces a less shattering effect.

A continuous process for manufacturing dynamite has recently been reported from Japan by the Asahi Kosei Kogyo Co. Dampened nitrocellulose is carried by conveyor belt to a spiked shredding wheel, and then allowed to fall down an inclined chute together with a stream of mixed glyceryl trinitrate and glycol trinitrate. The resulting slurry is in turn run on to a second conveyor belt which carries a controlled layer of finely powdered oxidizing agents such as sodium and ammonium nitrates. The gelatinizing mass is then chopped up by rotary knives and forced into a plastic tube. The dynamite dough is extruded from the tube by a system of rollers which exert a peristaltic squeeze, and can be loaded directly into cartridge cases.

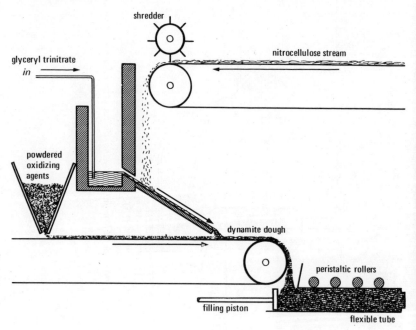

Fig. 4.4 Continuous manufacture of dynamite

Explosives

Pure glyceryl trinitrate is a sweet-tasting, dense, yellowish oil which freezes at 13·2°C to form rhombic crystals. It is soluble in most organic solvents but only slightly soluble in water. Although the vapour pressure is only 0·0015 mm Hg at 20°C, this is sufficient to produce severe headache (N.G. headache) in a sensitive subject by dilation of the blood vessels. Because of this effect it has been used medicinally as a vasodilator ('liquor trinitrini'). If ignited in the open it burns in small quantities but explodes when the temperature reaches about 220°C. It is very sensitive to shock—especially when heated or in the frozen state. The explosive force produced on detonation is due to the evolution of a large volume of gas which is further expanded by the high temperature attained during the reaction (3 400°C).

$$4 \begin{array}{c} CH_2ONO_2 \\ | \\ CHONO_2 \\ | \\ CH_2ONO_2 \end{array} \xrightarrow{\text{ignition or shock}} 12CO_2 \uparrow + 10H_2O \uparrow + 6N_2 \uparrow + O_2 \uparrow + 7 \cdot 23 \text{ megajoules}$$

Originally blackpowder was used as a propellant explosive in both small-bore weapons and cannon. It had the disadvantages, however, that it produced large volumes of smoke, rapidly fouled gun barrels, and deteriorated very quickly if allowed to become damp. As a result determined efforts were made in the mid-nineteenth century to discover new propellant explosives without these disadvantages. The most successful propellant resulting from this research was 'Cordite' which was discovered in 1888 by Abel and is still currently used for shells and bullets. This is manufactured by dissolving a mixture of GTN, mineral jelly and guncotton in acetone. The resulting dough is extruded to give macaroni-like strands which are then cut to length and dried. Other processes involve hot rolling the cordite dough or forming a slurry which can then be poured directly into a case and cured by heating.

In addition to guncotton and GTN, other nitric esters have also been used as commercial explosives. Although certain aliphatic nitric esters, such as methyl nitrate ($CH_3 \cdot ONO_2$), are powerful explosives, they are not so important commercially as the straight chain polyol nitric esters. Pentaerythritol tetranitrate (PETN) was prepared by Tollens in 1891 by nitrating pentaerythritol with concentrated nitric acid. Although a powerful explosive (95% as powerful as blasting gelatine) and chemically stable, its high cost has prevented it from competing with GTN in commerce, although it was extensively used during World War II as a bomb filling charge. It is also used as a filler for 'Cordtex' fuse and for making blasting caps.

$$\text{HOH}_2\text{C}-\underset{\underset{\text{CH}_2\text{OH}}{|}}{\overset{\overset{\text{CH}_2\text{OH}}{|}}{\text{C}}}-\text{CH}_2\text{OH} + 4\text{HNO}_3 \xrightarrow{+ \text{H}_2\text{SO}_4} \text{O}_2\text{NOH}_2\text{C}-\underset{\underset{\text{CH}_2\text{ONO}_2}{|}}{\overset{\overset{\text{CH}_2\text{ONO}_2}{|}}{\text{C}}}-\text{CH}_2\text{ONO}_2 + 4\text{H}_2\text{O}$$

pentaerythritol 'PETN'

Because of its sensitivity to friction and impact PETN is damped with water during storage and dried out immediately before use. It can be dispersed in molten TNT to produce explosives known as 'Pentolites', or mixed with a synthetic rubber or other plasticizer to give plastic explosives.

Large-scale production for the manufacture of plastics has reduced the price of PETN and could increase its importance.

Mannitol hexanitrate (MHN) is prepared by nitrating mannitol with concentrated nitric acid followed by precipitation with sulphuric acid. The first record of its preparation is by Domonte and Menart in 1847. Although not so chemically stable as PETN it resembles it in its explosive characteristics and is similarly limited in use by its high cost.

$$\underset{\underset{\text{CH}_2\text{OH}}{|}}{\overset{\overset{\text{CH}_2\text{OH}}{|}}{(\text{CHOH})_4}} + 6\text{HNO}_3 \xrightarrow{+ \text{H}_2\text{SO}_4} \underset{\underset{\text{CH}_2\text{ONO}_2}{|}}{\overset{\overset{\text{CH}_2\text{ONO}_2}{|}}{(\text{CHONO}_2)_4}} + 6\text{H}_2\text{O}$$

mannitol 'MHN'

Nitro-starch was first prepared by Braconnot in 1833 by dissolving starch in nitric acid and obtaining a white curdy precipitate of the nitric ester on pouring the product into excess water. It is still used as a commercial explosive on a small scale.

During World War I the Germans eked out their supplies of glycerol by adding up to 25% of sucrose (cane sugar) before nitration. Besides nitration of the sugar, however, a considerable degree of oxidation occurred with the generation of nitrous fumes and the consequent waste of nitric acid rendered the process uneconomic.

Explosives being used for quarrying. A short delay technique has been used in which a number of charges are detonated at very short intervals to produce a ripple of explosions along the quarry face

Explosive mixtures of ammonium nitrate (AN) with substances such as sawdust, naphthalene, picric acid, charcoal and dinitrobenzene were first described by the Swedes in 1867. This led to the use of ammonium nitrate as an explosive in its own right, and the manufacture of ammonia dynamites such as 'Monobel' (Du Pont).

Although difficult to detonate in small quantities, under pressure and when molten AN becomes much more sensitive. As a result, a number of disastrous explosions have occurred after fires involving large quantities of AN fertilizer. At Oppan (Germany) in 1921, 5 000 tonnes of AN loaded in railway trucks exploded killing over 1 000 people. During 1947 there were two ship explosions following fires involving AN fertilizer at Brest and Texas City causing widespread destruction and heavy loss of life.

In the early 1950's the Americans developed a form of AN known as 'prill' which would not cake or solidify on storage. This is manufactured by spray drying a hot concentrated aqueous solution of ammonium nitrate. The resulting beads are coated with diatomaceous earth dispersed in a wetting agent to facilitate its admixture with hydrocarbon oils. Although AN prill was originally used in 1955 in conjunction with carbon black ('Akremite'), more recently oil fuel, such as diesel oil, has been used for large-bore blasting operations. As the mixing is carried out on the spot this is commonly known as 'do-it-yourself' explosive. For instance, on the famous Mesabi iron ore range in Minnesota a 36 kg (80 lb) bag of AN is poured into a shot hole followed by 4·5 l (1 gallon) of diesel oil. This gives a cheap explosive with a force three times that of AN alone and is now termed ANFO (ammonium nitrate with fuel oil). Between 5·5 and 6% of diesel oil provides a balanced explosive with optimum detonative power and the minimum of nitrogen dioxide. In Canada mixtures of AN, TNT and water have been similarly used for blasting.

Two other types of explosive which are mixed immediately before use are the 'Sprengel' explosives patented in 1871 and the 'Lox' type explosives introduced by Linde in 1895. Sprengel explosives are mixtures of inflammable liquids, such as petroleum, carbon bisulphide and nitrobenzene, with powerful oxidizing agents, such as concentrated nitric acid and liquid nitrogen dioxide. The Hell Gate reefs which blocked the channel of the East River, New York, were blasted away with a Sprengel type explosive in 1885. In this instance nitrobenzene was poured over watertight cartridges containing potassium chlorate—a combination known as 'Rackarock'. Although powerful and cheap, the difficulties involved in the mixing and handling of these materials have prevented their extensive use.

Lox explosives are mixtures of liquid oxygen with carbon, or organic substances such as aniline and kerosine. Large-scale use of an explosive of this type ('Oxyliquit') was made during the construction of the Simplon tunnel in 1899 to avoid filling the workings with nitrous fumes. Disadvantages of Lox explosives are high sensitivity to spark and friction detonation and the rapid evaporation of the liquid oxygen after placing the explosive.

Explosives

The aromatic nitro compounds include a number of important commercial and military explosives. Unlike the nitric esters already mentioned the nitro group is in this case attached directly to a carbon atom.

$$\diagdown\!\!\!\!\!\diagup C-NO_2 \qquad \diagdown\!\!\!\!\!\diagup HCONO_2$$

nitro compound   nitric ester

A single nitro group, as in nitrobenzene, is insufficient by itself to produce an explosive substance, but two or more groups, as in trinitrobenzene, are effective. The preparation of aromatic substances with more than one nitro group is difficult. The nitro group has a powerful electron attracting effect which tends to deactivate the benzene ring. Thus in the case of nitrobenzene, electrophilic reagents can only attack the *m*-position and this with some difficulty. *m*-dinitrobenzene can only be prepared by nitrating nitrobenzene with a mixture of fuming nitric acid and concentrated sulphuric acid. The 1,3,5-trinitrobenzene derivative is even more difficult to prepare and the introduction of further nitro groups cannot be carried out directly.

nitrobenzene → (HNO₃ + oleum, >60°C) → *m*-dinitrobenzene → (prolonged nitration) → 1,3,5-trinitrobenzene

Trinitrobenzene (TNB) is a yellowish crystalline substance of great explosive strength, being the most powerful of all the trinitrated aromatic compounds. Despite its explosive nature it is chemically stable, but is difficult to prepare and has a melting point of 123°C which is inconveniently high for the filling of shells and bombs.

Trinitrotoluene (TNT or Triton) is much easier to prepare than TNB, the toluene used for its manufacture being extracted from coal tar or, since 1940, synthesized from heptane obtainable as a petroleum product.

Fuels, Explosives and Dyestuffs

toluene → (nitrating mixture) → 'TNT'

The first preparation of TNT was described by Wilbrand in 1863 but it was not used as an explosive for some 30 years. In World War I it became the most important of all the military explosives, being remarkably insensitive to shock and easily melted for pouring into shells, etc., as an explosive filling. It has an explosive power only about 65% that of blasting gelatine but, because of its widespread military use, it is often used as a standard for the 'yield' of military weapons. Thus atomic weapon yields are measured as equivalents of thousands of tonnes (kilotonnes) or millions of tonnes (megatonnes) of TNT. As it has only a small fraction of the oxygen required for complete combustion (negative oxygen balance), TNT burns with a very smoky flame. In explosive mixtures it is often combined with an ingredient such as ammonium nitrate which improves the oxygen balance and also cheapens the product. Thus Amatol, which is widely used, is a 60/40 mixture of TNT and ammonium nitrate.

Although insensitive to shock at ordinary temperatures, the sensitivity of TNT rises rapidly with temperature, especially under compression, and serious accidents have occurred through TNT igniting and then exploding with the rise in temperature.

Although originally produced by a batch process, TNT is now increasingly manufactured by continuous nitration. In the batch process nitrating acid (80% sulphuric acid/20% nitric acid) is slowly stirred into toluene, the temperature being maintained at about 70°C. When about half the required nitrating acid has been added the mixture is stirred for some hours. The remainder of the acid is then added, the temperature being allowed to rise to 110°C. Stirring is continued until nitration is complete, when the mixture is allowed to settle. While still hot, the spent acid is run off and used for preparing dinitrotoluene (DNT). The molten TNT is washed successively in hot water, sodium sulphite solution, very dilute sulphuric acid, and finally hot water. It is then allowed to solidify in shallow trays before breaking up into crumbs. The pure compound forms pale yellow crystals melting at 81°C, which are insoluble in water and very stable at ordinary temperatures.

An interesting nitro compound which is capable of being mixed with toluene or nitrotoluene to give an explosive mixture of great strength, is Tetan or tetranitromethane $C(NO_2)_4$, which was first produced in 1861 and is manufactured by treating acetylene with concentrated nitric acid. It has recently been used as a rocket fuel.

Before the widespread military use of TNT, trinitrophenol or picric acid was used as an explosive filling for shells, being first used by the French as Melinite in 1885 and later in the Russo-Japanese War of 1905. Meanwhile the British had produced a picric acid-based shell filling called Lyddite, so-called because it was first tried out on the artillery range at Lydd. This was used in the Boer War of 1899–1902. Although picric acid was superseded by TNT after about 1905, it was used by the Germans again in World War I because of the shortage of toluene. Dinitronaphthalene and hexanitrodiphenylamine were also used at this time and for the same reason. Also ammonium picrate has been used as a filling for armour-piercing shells because of its unusual insensitivity to shock.

Picric acid is a powerful explosive, intermediate in effect between TNT and TNB, which forms yellow crystals melting at 122°C, and which is sparingly soluble in water at room temperature. The metallic salts of picric acid are easily prepared and often form sensitive explosives. Great care must be taken, therefore, in the manufacture of the acid not to allow contact with metals. Lead-free paint must also be used for picric acid containers, since lead picrate is highly sensitive.

The lead salt of trinitroresorcinol, known as lead styphnate, is also used as a detonating explosive. It is usually prepared by stirring lead nitrate solution into a solution of magnesium styphnate and filtering off the red crystalline normal salt. One of the most useful properties of lead styphnate is its extreme sensitivity to electric sparks which makes it an ideal filling for electrically-fired detonators.

$$Pb(NO_3)_2 + \text{magnesium styphnate (aqueous)} \xrightarrow{60°C} \text{lead styphnate (aqueous)} + Mg(NO_3)_2$$

lead nitrate + magnesium styphnate (aqueous) → lead styphnate (aqueous) + magnesium nitrate

Compressed pellets of mono- and dinitroresorcinol are used as propellant fuels for toy rockets, model jet aircraft, etc. (Jetex).

Another sensitive trinitro aromatic compound is prepared by nitrating dimethylaniline and was first produced by Michler and Meyer in 1879.

dimethylaniline $\xrightarrow{\text{nitrating mixture}}$ 2,4,6-trinitrophenyl-methylnitramine

The pure compound is a colourless crystalline substance melting at 129°C, insoluble in water but readily soluble in acetone. It is rather unstable chemically and rapidly yellows on exposure to light.

It is commonly known as Tetryl although also termed Tetralite or CE ('compound explosive') and is used as a primer or detonating compound in anti-aircraft and high explosive shells. In recent years it has been largely replaced by PETN which is cheaper and more powerful.

The search for new high explosives still continues. Several novel compounds have been tested, including a highly thermostable Du Pont explosive which does not decompose even at temperatures of the order of 350°C. Nevertheless, none of these new materials has been a commercial success.

One of the best known components of detonating mixtures or initiating explosives is mercury fulminate, which has been used for this purpose since soon after its discovery by Howard in 1800, replacing the primitive flintlock ignition of blackpowder charges. It is a nitroso compound produced in the form of pale brownish needles by adding ethanol to a solution of concentrated nitric acid containing dissolved mercury. The reaction is carried out in large glass vessels ('balloons') because of the vigorous effervescence which takes place.

$C\equiv NOH$
fulminic acid

$\begin{matrix} C\equiv NO \\ \diagdown Hg \\ C\equiv NO \diagup \end{matrix}$
mercury fulminate

The fulminate forms an ideal detonating charge when mixed with 10–20% of an oxidizing agent such as potassium chlorate. It is very sensitive to shock or ignition and is stored in linen bags immersed in cold water in which it is insoluble. Decomposition occurs if the compound is not kept cool, however, and its deterioration in hot climates has led to its replacement by other initiators such as lead azide $(Pb(N_3)_2)$.

Lead azide is an example of an explosive compound containing weakly linked nitrogen atoms, capable of producing explosively large gas volumes on decomposition. Although less sensitive than mercury fulminate it has the advantages of being more stable and independent of the relatively scarce mercury. It was first used as an initiating explosive by the Prussian Government in 1893 but later abandoned as too dangerous. Interest was rekindled in its properties during World War I, however, due to the shortage of mercury for producing the fulminate. The danger of spontaneous detonation during manufacture can be prevented by co-precipitation with dextrin to prevent the formation of sharp-edged crystals and rigorous exclusion of metals which could form dangerously sensitive azides. For the same reason lead azide is sealed in aluminium caps and not copper ones, since copper azide is particularly sensitive to shock.

Lead azide is a salt of hydrazoic acid, which apart from being an unusual explosive compound in possessing no oxygen, has a most interesting structure. First prepared by Curtius in 1890 the free acid is a highly dangerous explosive and has caused a number of fatalities by accidental detonation during its preparation. It is only used technically therefore in the form of its salts. Although originally thought to have a cyclic structure it now seems probable that the nitrogen chain is linear.

$$H-N=N\equiv N$$
hydrazoic acid

$$\begin{array}{c} N\equiv N=N \\ \diagdown \\ Pb \\ \diagup \\ N\equiv N=N \end{array}$$
lead azide

An important self-linked nitrogen compound, which resembles somewhat the structure of TNT, is Cyclonite, which can be prepared by nitrating hexamine (hexamethylene tetramine) with fuming nitric acid. The reaction is carried out at around room temperature and the Cyclonite is precipitated as a white solid by chilling the solution to just above freezing point.

Cyclonite (RDX)

TNT

Although first discovered in 1899 it was not patented as an explosive until 1920 and only assumed importance as a bursting charge for shells and bombs during World War II. Because of the work carried out upon this material by the Research Department of Woolwich Arsenal it became known in this country as RDX. It has a superior shattering force (brisance) to PETN but is a little more expensive to produce and rather too sensitive to impact for convenience, unless desensitized with wax or some other inert filler. A mixture of Cyclonite with TNT and aluminium powder was used in World War II as an underwater explosive to combat submarines, under the name of Torpex. Cyclonite also forms a powerful explosive mixture with ammonium nitrate. A similar compound to RDX may be prepared by nitrating hexamine dinitrate in the presence of acetic anhydride. The resulting product (HMX) is a stable explosive of high yield with the following structure:

$$\begin{array}{c} \text{O}_2\text{N} \diagdown \quad \overset{\text{H}_2}{\text{C}} \quad \diagup \text{NO}_2 \\ \diagdown \text{N} \quad \text{N} \diagup \\ \text{H}_2\text{C} \diagdown \qquad \diagup \text{CH}_2 \\ \diagup \text{N} \quad \text{N} \diagdown \\ \text{O}_2\text{N} \diagup \quad \underset{\text{H}_2}{\text{C}} \quad \diagdown \text{NO}_2 \end{array} \quad \text{tetramethylene-tetranitramine} \quad \text{(HMX)}$$

The favoured high explosives used for military purposes are RDX, TNT and Amatol. Because of their insensitivity a small primary charge or 'gaine' is usually used in addition to the detonator to ensure satisfactory performance. The gaine commonly consists of a narrow metal tube packed with Tetryl pellets. Shells and grenades have relatively thick casings which fragment on explosion, while bombs and depth charges which rely upon the production of shock waves have relatively thin casings.

The diazo compound, diazo dinitro phenol, known as DDNP or Dinol, is of historical interest because it was discovered in 1858 by Peter Griess, who was later to discover the diazo dyestuffs. It is also of technical interest because of its powerful explosive properties, and is widely used in the USA as a detonating explosive.

SPECIAL USES OF EXPLOSIVES

The coal mining industry in this country is easily the largest user of explosives, still consuming about 25 000 tonnes annually, each tonne of coal mined requiring about 0·1 kg (¼ lb) of explosive on average. Early use of gunpowder in coal mining was fraught with great danger because of the hazard of accidental detonation of the 'firedamp' present in ill-ventilated workings. Indeed many pit disasters in the past were the result of using explosives to bring down coal. In the seventeenth century a 'fireman' covered with a wet cloth used to crawl into the workings each morning and ignite the gases with a candle on a long pole! Despite this, violent explosions frequently took place, such as that at Gateshead in 1705 when several miners were blown right out of the pit—'Slain in a coal-pitt in the Stony Flatt, which did fire', as the event is described in the parish register. During the latter part of the nineteenth century great efforts were made to overcome the danger of using explosives underground and a number of commissions deliberated the problem in France, Germany and Britain.

As a result the British Coal Mines Regulations Act of 1887 prohibited the use of explosives in coal mines except under certain conditions, and steps were taken to minimize the flame produced by explosions. Thus gunpowder was pelleted and covered in a layer of paraffin wax to form a coolant sheath and to allow water quenching of the flame. This type of treated explosive was known as 'Bobbinite' and was in use in this country until the 1930's.

The Coal Mines Regulations Act was revised in 1896 and a list of 'Permitted' explosives (now known as 'P 1 explosives') was drawn up, including the only types to be used in coal mining. Simultaneously a test gallery was built at Woolwich Arsenal in order that mining explosives could be tested officially. This testing device is still used today, but has been situated at Buxton since 1929. It consists in essence of the firing of charges of the explosive under test from a small cannon into a long cylindrical metal tunnel (gallery) containing suspensions of coal dust or explosive mixtures of air and methane.

The first attempts to produce a safe mining explosive involved the addition of cooling agents such as salt, borax and sodium bicarbonate to the charge. In this way the explosion flame could be cooled below the ignition temperature of the firedamp without seriously reducing the force of the explosion. In 1914, however, the Belgian chemist Lemaire introduced the idea of sheathed ('P 2')

Test galleries at the Safety in Mines Research Establishment (Sheffield). *Above:* The cannons used for firing the explosive charges under test are being held in position against the ends of the large galleries which contain air mixed with methane or coal dust. *Below:* A test charge being inserted in the cannon before firing in the gallery on the left

explosives. These had an outer sheath of inert cooling agent surrounding the explosive charge instead of a mixture of the two ingredients. This type of explosive was introduced into Britain in 1934, the usual coolant being sodium bicarbonate, which blanketed the explosion with a mixture of steam and carbon dioxide. Although sheathed explosives had a good safety record they had the serious disadvantage that the sheath might be accidentally or deliberately removed. To avoid this hazard a new type of mining explosive was introduced in 1949. This was designated as a 'P 3' or 'Eq.S.' explosive ('equivalent-to-sheathed') indicating that it was as safe to use as a sheathed explosive. The Eq.S. explosives are improved forms of the original 'cooled' explosives in which an inert component is included in the formulation to quench the explosion flame. Salt is still widely used as a coolant for this purpose with explosive mixtures containing low freeze nitroglycerine, with nitrocellulose, TNT, or ammonium nitrate. These explosives can be effectively waterproofed by the addition of about 1% calcium stearate, or more recently sodium carboxy methyl cellulose which swells up on contact with water and seals off the charge. A problem arises in the formulation of P 3 gelatine explosives because the addition of salt tends to affect their consistency. This has been resolved by substituting salt for part of the ammonium nitrate content.

The Eq.S. gelatine explosives have proved very successful and since the first one ('Unigel') appeared in 1954 they have been widely adopted and several types are now available ('Denespex', 'Pentregel'). The rising popularity of the Eq.S. permitted explosives is shown in Fig. 4.5.

After the introduction of Eq.S explosives, the possibility of accidental firedamp explosions as a direct result of the use of explosives became remote, except in the case of misuse or under hazardous conditions. Two dangers are potentially explosive suspensions of fine coal-dust in the atmosphere, and the presence of gas-filled fissures in the coal. In an attempt to eliminate these hazards the idea of 'pulsed infusion' was introduced, which is a combination of shot-firing and the older technique of water infusion, i.e. water at high pressure being forced into the cracks in the working face to bring down the coal. The primed charge is placed in the shot hole and then water at high pressure applied.

Explosives

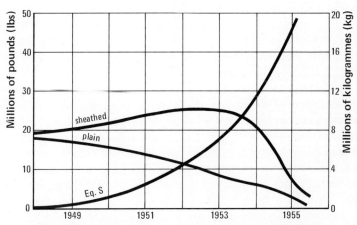

Fig. 4.5 Consumption of permitted explosives, 1944–1955

Explosion of the charge sets up very high pressures in the water-filled hole and opens up the fine cracks in the coal face, thus breaking up the coal without producing dust. In addition the water blankets the charge and is forced into any fissures which might contain gas.

The explosive used for this purpose has to function properly under high pressures of water which may reach $7·75$ MN/m^2 ($\frac{1}{2}$ tonne/

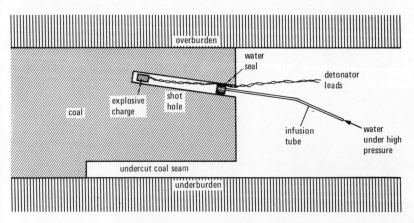

Fig. 4.6 Pulsed infusion blasting

in^2) pressure. This is ensured by incorporating a small quantity (3–10%) of barium sulphate in a nitroglycerine explosive ('Aquaspex', 'Hydrobel'). The barytes, which is finely powdered, is readily wetted by the nitroglycerine and has the unusual property of sensitizing it under the special conditions involved.

Explosive charges used for seismic geological exploration either underground or below the sea have to meet the same exacting requirements as for pulsed infusion shot-firing, and therefore have similar formulations. Imperial Chemical Industries have produced a range of explosives for this purpose known as 'Geophexes' which contain powdered barium sulphate and have been successfully detonated at sea depths of up to 830 m (2 750 ft). Explosives of this type were used by Sir Vivian Fuchs during his Transantarctic Expedition to find the thickness of the polar ice-caps.

Ammonium nitrate explosives hermetically sealed in metal canisters have also been used for seismic purposes. With established marine prospecting systems larger seismic charges of about 23 kg (50 lb) of explosive are used. A large quantity of the explosive energy produced, however, is lost in the resultant gas bubbles, the oscillation of which often interferes with the seismic record. Experiments showed that if the explosives were in the form of a thin strand, bubble formation was minimized and greatly improved acoustic effects were produced. This has led to the introduction by ICI of a line charge termed 'Aquaflex Seismic Cord' consisting of a core of pentaerythritol tetranitrate (PETN) within a plastic sheath. This is used in 30 m (100 ft) coils representing an explosive charge of 0·7 kg (1½ lb), producing a result comparable to that obtained by the 23 kg (50 lb) conventional charge.

Pulsed infusion shot-firing. The primer cartridge is being drawn in to the shot hole. Note the Davy lamp for detecting the presence of 'firedamp' (methane)

Explosives

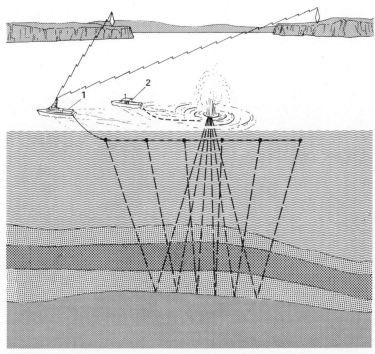

Fig. 4.7 Underwater seismic survey: 1) boat towing seismic recorders—position fixed by radio beacons; 2) boat for setting and detonating explosive charges

The use of 'shaped' explosive charges was proposed as early as 1888 by Munroe. Using shaped metal cones or tube structures in front of an explosive charge has the effect of concentrating the blast from the explosion into a narrow and exceedingly powerful jet. This is useful when the explosion is intended to penetrate a barrier of some kind. Although at first regarded rather as novelties, the success in World War II of the 'bazooka' and 'PIAT' anti-tank weapons, which used shaped charges to penetrate armour plate, led to several commercial applications after the war. Examples of peacetime applications of shaped charges are the perforation of oil well casings and the piercing of the tapping plugs on open hearth steel furnaces.

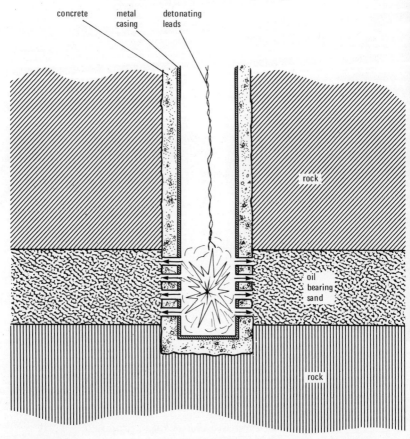

Fig. 4.8 Use of shaped charge perforator to pierce oil well casing

Specially shaped wedge charges have also been used to cut underwater telegraph cables to facilitate lifting for repair or inspection.

Another interesting commercial application of explosives is in the shaping or engraving of metal. Complex pressed shapes can be produced by exploding a shaped charge above a metal sheet pressed against a mould. Similarly, intricate patterns can be produced on a metal surface by covering it with a stencil and exploding a charge above it. Another commercial application of explosives is in the

Above: Releasing a dynamite charge during a seismic survey off the Nigerian coast. The charge is suspended from the balloon at the correct depth and then detonated by radio

Right: Blasting explosives being cartridged in ICI's Ardeer factory, Ayrshire. This factory was founded by Alfred Nobel to make dynamite, the first effective commercial explosive based on nitroglycerine

production of clad metal sheet. A thin sheet of cladding metal is inclined at an angle to the firmly fixed base material. On firing a small explosive charge immediately above the cladding, the latter is welded to the base.

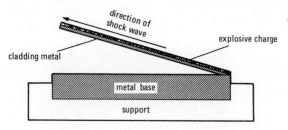

Fig. 4.9 Explosion cladding technique

Explosives have also been successfully used to imbed reinforcing filaments into a metal matrix. Sheet copper has been reinforced with fine tungsten wire in this way, producing a composite bar of great strength and stiffness. A sandwich of wire and metal sheets is built up on a metal anvil and covered with a sheet of explosive. The explosion forces the layers together into a composite structure whilst retaining the identity of the reinforcing wires.

For rocket propulsion composite propellants are often used which consist of a powerful oxidizing agent such as ammonium perchlorate bound with a plastic fuel (e.g. poly-iso-butene) to give a rubbery solid. Nitrocellulose/nitroglycerine mixtures such as the American rocket propellant JPN are also used. To achieve a steady rate of burning the propellant charges of rockets are tubular. The inner surface is convoluted so that the area of combustion remains virtually unchanged.

One of the most remarkable of all rocket fuels is Monex W which has been recently developed for use in manoeuvring spacecraft. This is based upon human sewage and other waste material to which is added powdered aluminium and ammonium nitrate. Not only is Monex W claimed to be an effective propellant but it affords a novel way of ridding manned spaceships of accumulated organic debris.

Another interesting use of propellant explosives is in the manufacture of starter cartridges for jet engines. These contain a hollow charge in a metal case which is closed at one end by a plastic disc. On detonation a large volume of hot gas is produced which passes through the turbine and compressor of the engine, enabling it to be started rapidly. Smaller cartridges of a similar type are used in special guns to drive nails or bolts into masonry. Bolt guns also have been used to drive lifting hooks into submerged wrecks.

Chapter 5
Dyestuffs

> The Liquors that Dyers employ to tinge are qualified to do so by multitudes of little corpuscles of the Pigment or Dyeing stuff, which are dissolved and extracted by the Liquor, and swim to and fro in it, these Corpuscles of Colour (as the Atomists call them) insinuating themselves into, and filling all the Pores of the Body to be Dyed.
>
> Robert Boyle

The dyeing of textiles to give them a 'fast' colour difficult to remove by washing or light bleaching was, for many centuries, carried out using organic dyes extracted from plant material (indigo), and occasionally from animals such as insects (cochineal) or molluscs (Tyrian Purple). As early as 3 000 BC the Egyptians and Chinese were dyeing fabrics, as shown by pictures and other relics which have been discovered. The Indians and Persians were using plant materials to dye silks, brocades and wool before 2 000 BC and by about 1 400 BC the Egyptians were producing fine dyed cloth using materials such as Safflower (Dyers Thistle) to give a wide range of colours.

In Roman times Tyrian Purple was extracted from a Mediterranean mollusc (Murex) and the rich colour obtained was only used by persons of the highest rank. Lucretius refers to the fastness of this dyestuff in his *De Rerum Natura*:

'The purple dye of the shellfish so unites with the body of wool alone, that it cannot in any case be severed—not if the whole sea were willed to wash it out with all its waters.'

Pliny also describes many dyeing recipes used in Roman times, and excavations in the ruins of Pompeii revealed a dyer's shop decorated with murals depicting current dyeing techniques. During Roman times it was also known that a variety of indigo called woad was being used by the ancient Britons.

Dyestuffs

After the fall of Rome there is little record of dyeing being practised until the formation of the Dyers Guilds in the fourteenth and fifteenth centuries which resulted in a great improvement in the standard of dyeing and control of dyeing processes. Many of the dyes used at this time could only be used satisfactorily on fibres which had been mordanted (Fr. *mordre*—to bite) by pre-treatment with aluminium, iron or tin salts. Indigo, alizarin ('madder' or Turkey Red), log-wood and cochineal were commonly used as mordant dyes, and are still occasionally used today.

In 1856, however, a discovery was made which was to revolutionize the practice of dyeing and stimulate the whole field of chemical industry.

William Perkin attended the City of London school where he first became interested in chemistry. At the age of fifteen he obtained a place at the Royal College of Chemistry (later to become the Imperial College of Science and Technology—South Kensington) which had recently been founded under the direction of the distinguished German chemist Hofmann. By the age of seventeen this son of a London builder had completed his course and was appointed as a junior assistant. During the Easter vacation of 1856 Perkin attempted the synthesis of quinine ($C_{20}H_{24}N_2O_2$) from alkyl toluidine ($C_{10}H_{13}N$) by oxidation, working in an improvised laboratory at his home in Norwood. The experiment was pre-destined to failure, since, unknown to Perkin, the structures of the two materials were unrelated.

The reddish precipitate he obtained interested Perkin, however, and he continued his experiments by oxidizing other amines. On treating aniline sulphate with a mixture of sulphuric acid and potassium dichromate he obtained a black precipitate from which he was able to extract a purple dye using ethanol. Encouraged by his friend Church, and Thomas Keith, a silk dyer from Bethnal Green, Perkin sent a sample of silk dyed with this new material (Mauve) to the firm of Pullar's of Perth. So enthusiastic were the reports received by Perkin that with the help of his brother and father he began building a factory to manufacture this new material in June 1857 at Greenford Green. With no previous experience this young man not only solved the many problems associated with the design of the required chemical plant, but pioneered the commercial preparation of aniline by the reduction of nitrobenzene. In addition to this he developed techniques for carrying out the dyeing of silk and cotton, some of which are still in use.

The new colour became very popular during the late Victorian era, especially after the Queen had appeared in public in 1862 wearing a frock coloured with the new dye. Even the penny stamps were printed in mauve from 1881 to 1901 ('penny lilacs'). The outstanding success of Perkin's Purple not only stimulated chemists into searching for other dyes based upon aniline and related coal tar derivatives, but laid the foundation of the organic chemical industry.

The feverish activity in the field of synthetic coal tar chemistry was reflected in a poem which appeared in a contemporary issue of *Punch*, and which concluded as follows:

> Oil and ointment, and wax and wine,
> And the lovely colours called aniline,
> You can make anything, from a salve to a star,
> (If you only know how), from black coal tar.

During 1859 the French chemist Verguin discovered the first of a brilliant series of triphenylmethane dyestuffs, which he christened Magenta in commemoration of a recent victory by the French against the Austrians. Another interesting development resulted from attempts by Nicholson in 1862 to solubilize a violet dye (Bleu de Lyons) produced on heating magenta with aniline. He discovered that sulphonation of the dye by treatment with sulphuric acid rendered it water soluble—a technique which is now widely used. Perkin's old master Hofmann also produced a series of brilliant dyes by methylating or ethylating the amino groups of the magenta molecule.

In this way Methyl Violet was discovered, which, in addition to its intense colour, possessed interesting therapeutic properties. In 1868 Liebermann and Graebe in Germany synthesized alizarin from the coal tar derivative anthracene. As alizarin is the coloured component of the madder plant this was the first naturally occurring coloured material to be synthesized. By the following year Perkin had worked out a new synthesis of alizarin suitable for large-scale production and increased his output from 1 tonne in 1869 to 435

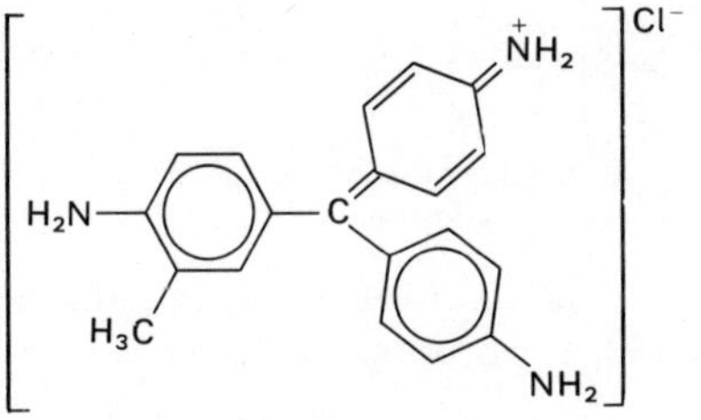

Magenta

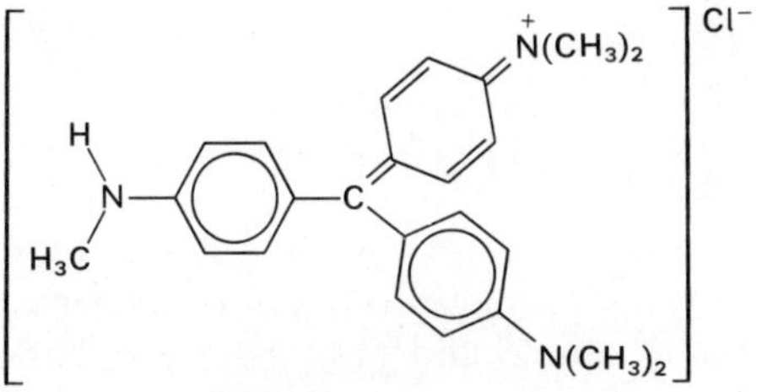

Methyl Violet

tonnes in 1873. The German dyestuff industry was not only able to match Perkin's output but doubled it, and in 1874 Perkin gave up the struggle and sold his factory. With their main competitor out of the field the Germans quickly captured most of the market and it is estimated that by 1885 80% of the dyes bought in this country were of German origin.

Meanwhile in 1858 a young German chemist, Peter Griess from Marburg University, was working on an interesting new discovery in the laboratories of the Ind Coope brewery at Burton-on-Trent. He was able to isolate entirely novel types of chemical compound from the reaction of nitrous acid and aromatic primary amines. These unstable, easily decomposable materials were termed diazo or diazonium compounds as they were thought to contain a pair of double bonded nitrogen atoms (—N=N—), although recent work shows them to possess an ionic structure resembling ammonium salts. Research on the diazo compounds was greatly facilitated by the discovery by Martius in 1866 that instead of using nitrous acid, which was difficult to prepare and handle, a mixture of sodium nitrite and dilute hydrochloric acid could be used, forming the corresponding diazonium salt.

Fuels, Explosives and Dyestuffs

aniline → benzene diazonium chloride (NaNO₂ + HCl, 0°C)

In 1859 Griess prepared the first azo dyestuff by linking aniline with a diazonium salt (coupling) to produce Aniline Yellow (aminoazobenzene).

benzene diazonium chloride + aniline → aminoazobenzene (Aniline Yellow)

A spate of azo dyes now began to appear when it was realized that diazonium salts could be coupled with a wide variety of aromatic amines and phenols to give basic and aniline azo dyes respectively. Originally the basic azo dyes were of most importance, with the discovery of Bismarck Brown by Martins in 1863 and Chrysoidine G by Caro in 1875. Both of these are still used to dye leather and paper. Bismarck Brown is of special interest because it is a diazo (or tetra-azo) compound, the use of a double diazonium salt enabling two aryl groups to be coupled simultaneously.

In 1876 one of the first acid azo dyes Orange II was produced by coupling diazotized sulphanilic acid and β-naphthol. These acid dyes soon became popular for dyeing wool and silk, the first real commercial success being Fast Red AV prepared by Caro in 1877 from two naphthalene intermediates. The success of Fast Red led to the experimental coupling of a number of other naphthols and diazotized naphthylamines, producing other shades of red such as Carmoisine (1883). By this time it was common practice to use sulphonated coupling intermediates to increase the solubility of the azo dye—an idea suggested by Nicholson, another pupil of the great Hofmann.

Up to this time, although many of the azo dyes produced had an affinity for wool and silk, which were proteins, little success had been experienced in dyeing cellulosic fibres, such as cotton, even after mordanting. In 1884, however, Congo Red was produced by Böttiger as the result of coupling sulphonated α-naphthylamine with the tetra-azotized double-ringed compound benzidine.

Sampling a batch of azo dyestuff, prior to isolation by filtration

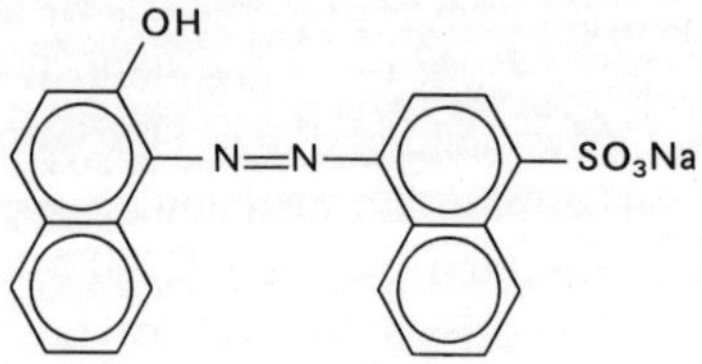

Fast Red AV

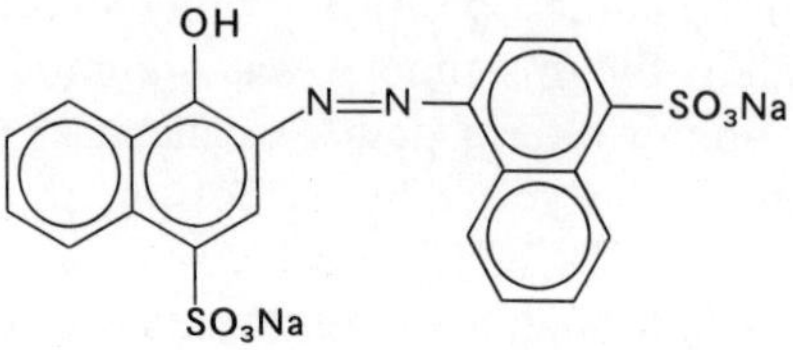

Carmoisine

It was found that Congo Red would dye unmordanted cotton at boiling point, and dyes of this type were termed 'direct' dyes. A complete range of direct cotton dyes was later produced by Bayer and other German firms by building different structures on the original Congo Red framework.

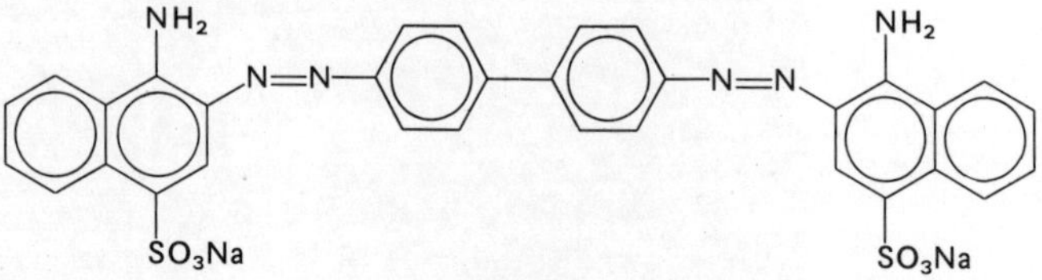

Congo Red

One outstanding disadvantage of the direct dyes, however, was their poor resistance to washing, and intensive research was carried out to make the direct dyes 'fast' to laundering and sunlight. One method was to form a metal compound by treating the dyestuff with a solution of a salt of iron, chromium, cobalt or copper (mordant dyes). This linked together a cluster of dye molecules and increased the fastness of the dye. Unfortunately this technique could only be used with molecules having adjacent hydroxyl (—OH) groups to which the metal atoms could be coupled, as in the case of the following blue azo dye.

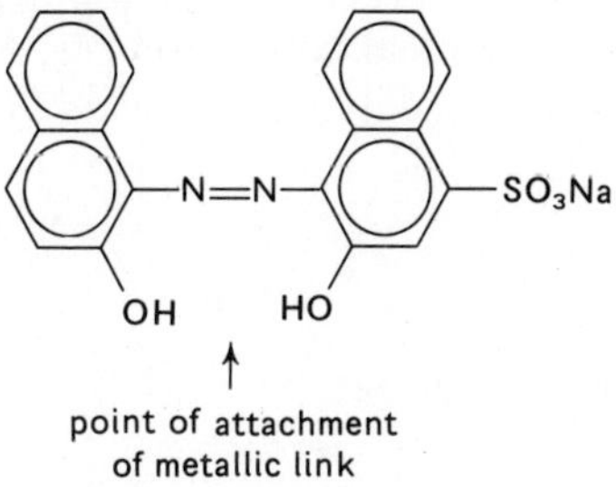

[Structure: naphthol-N=N-naphthol-SO₃Na with OH groups, showing point of attachment of metallic link]

point of attachment
of metallic link

A more successful idea was first demonstrated by Green using the yellow dye Primuline. Cotton could be dyed directly with this material but the fastness was very disappointing. Since the Primuline molecule had a suitable amino group ($-NH_2$) available, however, Green then diazotized the dyestuff by treating the dyed cotton with nitrous acid. He was then able to couple the diazonium compound produced with β-naphthol to give an insoluble dye—the azo dye being formed on the fabric itself (ingrain or developed dyeing). The method proved to be commercially successful, and in 1885 Para Red was introduced by the Badische Anilin und Soda-Fabrik company in Germany, as a developed cotton dye. The cotton fabric was first impregnated ('padded') with β-naphthol and then treated with a solution of diazotized p-nitro-aniline. Para Red soon replaced alizarin as a fast cotton dye, although the lack of affinity of β-naphthol to cellulose created manufacturing difficulties only overcome in 1912 when the β-naphthol was replaced by the anilide of β-oxynaphthoic acid, which is strongly attracted to cellulose.

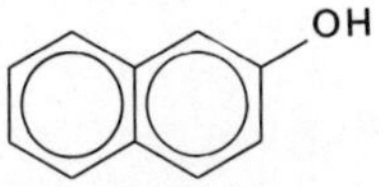

β-naphthol
(2-naphthol)

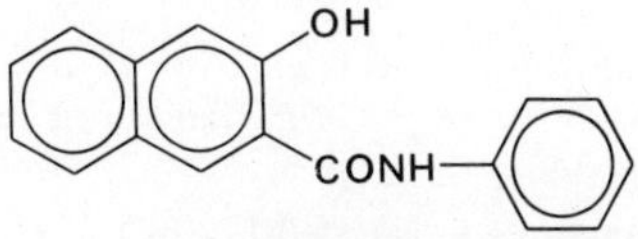

β-oxynaphthoic
acid anilide

The 1880's also finally saw the unravelling of the mystery which shrouded the structure of indigotin—the natural colouring matter present in the indigo plant and woad. The first step carried out in 1826 by Unverdorben was the production of aniline from natural indigo by heating, later workers identifying first anthranilic acid and then isatin in aniline breakdown products. At this point Bayer set himself the task of elucidating the structure of indigotin, which he succeeded in doing in 1883.

(a) aniline

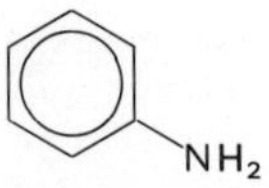

(b) anthranilic acid

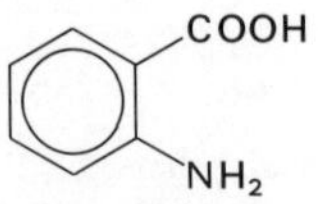

(c) isatin

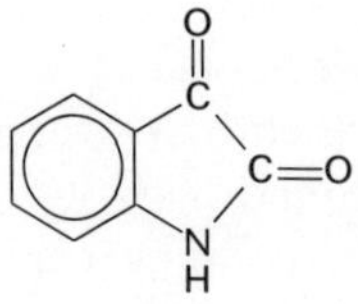

(d) indigotin

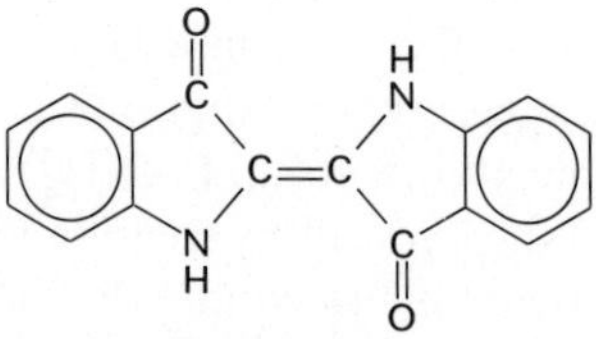

Seven years later a synthesis of indigotin from phenylglycine was worked out by Heumann and adopted the following year by the Badische company. Tyrian Purple is a dibromo derivative of indigotin—an interesting discovery made at the turn of the century by Friedländer.

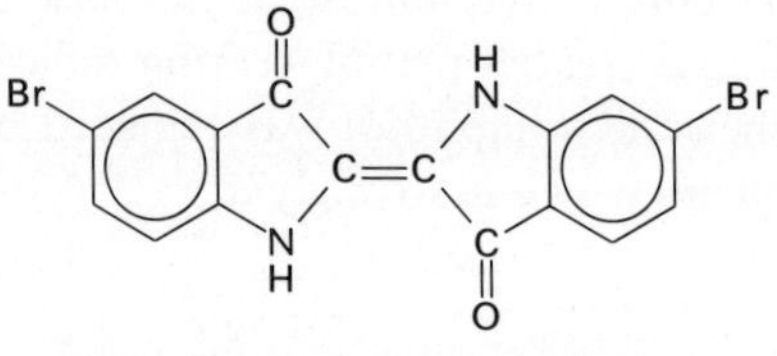

Tyrian Purple (dibromoindigotin)

Although insoluble, indigotin can be converted to the soluble pale green dihydro derivative by treating with a reducing agent such as alkaline sodium dithionate ($Na_2S_2O_4$). This is known as the 'leuco' (Gk. *leukos*—white) form of indigotin and is soluble in alkaline solution. It also has an affinity for cellulosic and protein

Dyestuffs

fibres, and is readily re-oxidized back to the deeply coloured insoluble indigotin. By steeping fabric in a solution of the leuco form of the dye, and then exposing the impregnated material to the atmosphere or treating with an oxidizing agent, an extremely fast colour is produced. This is known as vat dyeing and although rather a troublesome process the excellent fastness of the product led to the development of other vat dyes based upon the indigotin framework, such as tetrachloroindigotin ('Ciba Blue BR'), and more recently 'Ciba Green G' and 'Indanthrene Blue 2G'. The latter has found considerable application as a printing dye and is produced by coupling isatin with an α-naphthol derivative to produce a kind of 'half' indigotin molecule.

(a) Ciba Blue BR (tetrachlorindigotin)

(b) Indanthrene Blue 2G

An interesting development in the field of vat dyes was the discovery in 1901 by René Bohn that an indigotin-like dye, Indanthrone could be prepared by fusing β-aminoanthraquinone with caustic potash in the presence of phenol.

β-aminoanthraquinone $\xrightarrow{KOH/KClO_3, 200-250°C}$ Indanthrone

This was the first of a number of dyes based upon anthraquinone and known as quinonoid dyes. Some of the quinonoid dyes have large molecules containing multiple ring structures which increase their affinity for cellulosic and protein fibres. Examples of these are found in the ICI 'Caledon' range.

Caledon Yellow 4G

In 1921 it was discovered that a soluble form of vat dye could be prepared from the disulphuric ester of the leuco form, converting this into the sodium salt by treating it with sodium hydroxide. The resulting compounds, which are stable, soluble in water and have an affinity for cellulosic fibres, are termed Indigosols and greatly simplify the vat dyeing process.

Another type of cotton dye was introduced by the French chemist Raymond Vidal in 1893. These were characterized by the inclusion of sulphur linkages in their molecular framework and were for this reason termed sulphur dyes. Although fast to light and water, the sulphur dyes never achieved great commercial importance because of their rather dull colours and their sensitivity to polluted city air.

In contrast to this the 'reactive' 'Procion' dyes first produced by ICI in 1965 and the 'Cibacron' dyes produced in the following year by the Swiss Ciba Company, represented a major advance in dye technology. Ratee and Stephen working in the research laboratories of ICI Dyestuffs Division discovered that the cyanogen chloride trimer cyanuric chloride could be incorporated into a dyestuff without interfering with its colour. Since cyanuric chloride also reacts with cellulosic, protein and nylon fibres, it can act as a bridging link between dye and textile. The novel feature of the reactive dyes is their ability to react chemically with the fibre molecules to produce a coloured compound. Thus with cotton an ether linkage is formed

Reaction vessels used in the manufacture of 'Procion' reactive dyes

Multi-purpose plant used in the manufacture of dyestuff intermediates

by the interaction of one of the active chlorine atoms of the cyanuric chloride with one of the cellulose hydroxy groups.

(a) cyanogen chloride → (polymerization) → cyanuric chloride

(b) cyanuric chloride + R—NH—H (coloured amine compound) → (−HCl) → dye complex (where R is an aromatic group)

(c) dye complex + HO—CH (part of cellulose molecule) → (NaOH) → dye complex attached to cellulose by ether linkage

Reactive azo dyes can be prepared in a similar way by treating aminonaphthol sulphonic acids with cyanuric chloride and coupling the complex formed with a diazonium compound.

A new type of pigment was accidentally produced during the manufacture of phthalimide at the works of Scottish Dyes Ltd in 1928. An intense blue colour, produced by the action of ammonia and molten phthalic anhydride on the exposed iron of a damaged reaction vessel, was identified in 1934 as a complex of iron and phthalocyanin (literally 'blue phthalic compound'). The phthalocyanins are large ring ('macrocyclic') structures produced by the linking together of four phthalonitrile molecules around a central metal atom. The natural pigments termed porphyrins, such as haemoglobins and chlorophyll, have similar structures.

The colour of the phthalocyanins is dictated by the nature of the metal atom present. Copper phthalocyanin, first marketed by ICI in 1934 as 'Monastral Fast Blue BS', is commercially of greatest

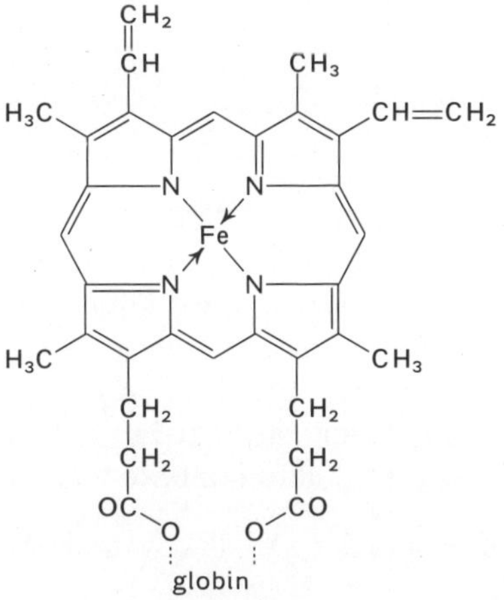

haemoglobin

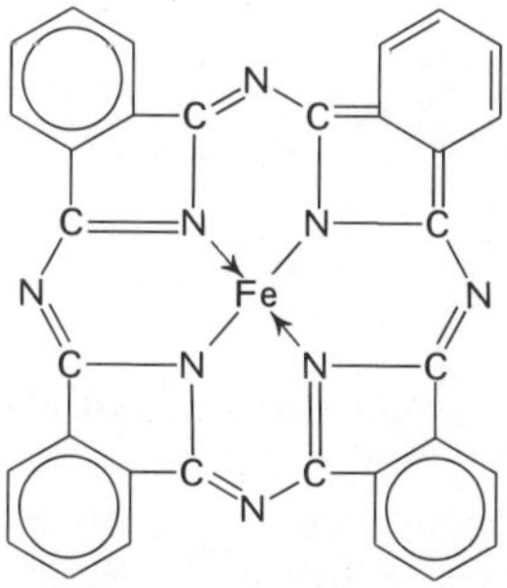

iron phthalocyanin

importance. Being i̶n̶s̶o̶l̶u̶b̶l̶e̶, the phthalocyanins cannot be used as fabric dyes, but they a̶r̶e̶ ̶i̶n̶ ̶d̶emand as pigments for colouring plastic, paper, paints and ink. They are extremely stable and exceptionally resistant to light bleaching.

Cyanine dyes were discovered by Vogel in 1873 to have the unusual property of sensitizing photographic emulsions to light at the red end of the spectrum. The parent dye cyanine (Quinoline Blue) was of little commercial use as it caused fogging, but the later carbocyanines which contained two heterocyclic nuclei joined by a conjugated hydrocarbon chain were to make an important contribution to colour and infrared photography. Early compounds such as Sensitol Red (1905) and Kryptocyanine (1917) possessed a short conjugated chain of four carbon atoms, but later it was discovered that by lengthening the chain increased sensitivity to long wavelength light was obtained. The polycarbocyanines with up to 11 conjugated chain carbons sensitize photographic emulsions into the far infrared.

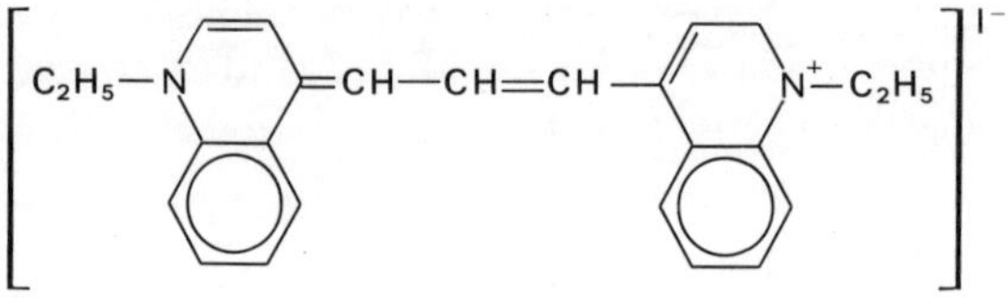

$$\left[C_2H_5-N \underset{}{\bigcirc}\!=\!CH-CH\!=\!CH-\underset{}{\bigcirc}\!\overset{+}{N}-C_2H_5 \right] I^-$$

Kryptocyanine

COLOUR AND CHEMICAL CONSTITUTION

Liebermann and Graebe suggested in 1868 that since the colour of organic dyestuffs was often destroyed by reducing bleaches the phenomenon of colour was associated with unsaturation. A decade later Witt put forward the theory that colour was linked to special colour promoting radicals which he termed chromophores. These included the carbon/carbon double bond ($>\!C\!=\!C\!<$), and the carbonyl ($>\!C\!=\!O$), nitro ($-\overset{+}{N}\!\!<\!\!\overset{O}{\underset{O}{}}$), nitrile ($-C\!\equiv\!N$), diazo ($-N\!=\!N-$) and nitroso ($-N\!=\!O$) groups to which was later added the quinonoid grouping ($=\!\!\bigcirc\!\!=$). Compounds containing chromophores were termed by Witt chromogens, although they were not necessarily coloured. He also pointed out that other groups acted in an auxiliary capacity by imparting colour to colourless chromogens and intensifying the colour of others. These were termed auxochromes and included amino ($-NH_2$) and substituted amino groups, hydroxy ($-OH$) and alkoxy groups ($-OCH_3$, $-OC_2H_5$). The possession of auxochromes also rendered a dyestuff more soluble and increased its affinity to textile fibres. The role of chromophores and auxochromes according to Witt's theory is demonstrated by means of the examples shown opposite.

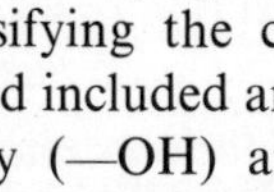

The darkening of vegetables and fruits such as potatoes and apples when cut open and exposed to the air is due to oxidation of

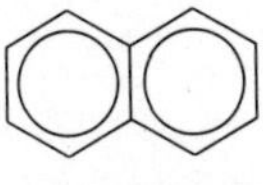

naphthalene
(no colour)

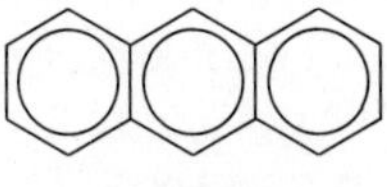

anthracene
(pale blue fluorescence)

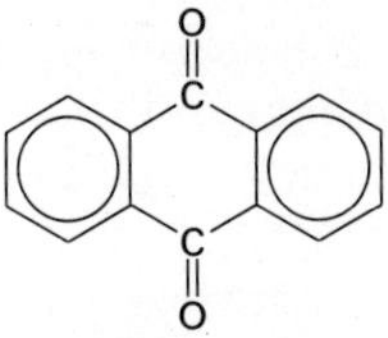

anthraquinone
(bright yellow)
chromophore

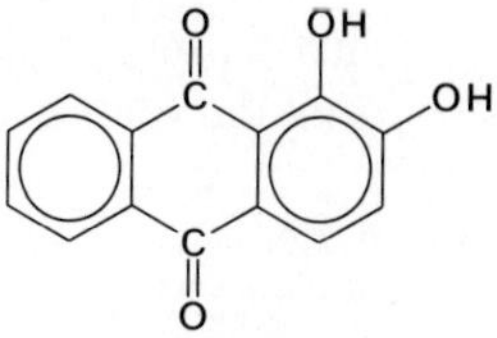

alizarin
(ruby red)
chromophore + auxochrome

the polyphenol catechol, firstly to the corresponding quinone and then to the hydroxyquinone which polymerizes to form dark melanic-type pigments.

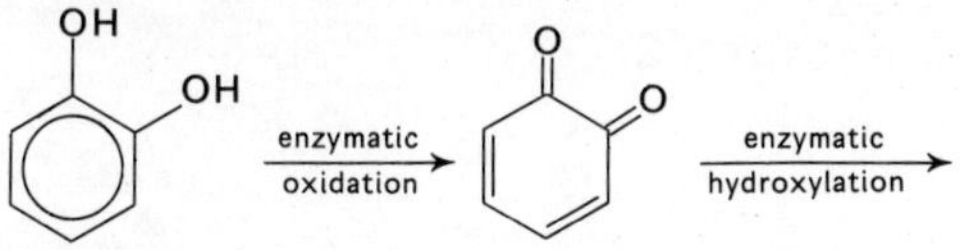

catechol →[enzymatic oxidation] o-quinone →[enzymatic hydroxylation]

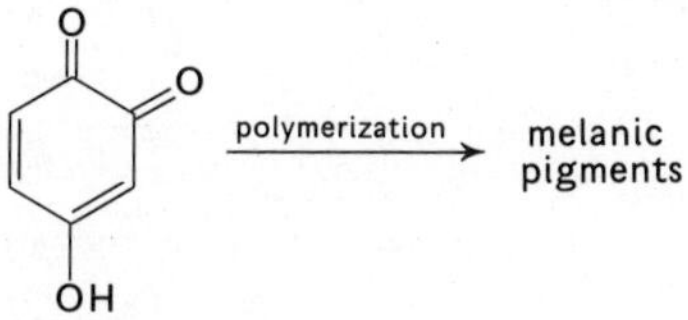

hydroxyquinone derivative →[polymerization] melanic pigments

Recently fresh evidence on the relation between colour and chemical constitution has been promoted by ultraviolet and visible spectroscopy. This, together with our greater understanding of chemical structure since the application of quantum and wave mechanics, has led to modification of these early theories.

It has been shown that all organic substances absorb radiant energy, especially in the high energy short-wave region of the electro-magnetic spectrum. Only if the absorbed radiant energy occurs in the small visible range 400–750 nm (4 000 Å–7 500 Å) does a molecule appear coloured, the colour perceived being complementary to that absorbed. Thus a substance which absorbs light at the blue end of the spectrum appears reddish-yellow while one absorbing the whole visible range appears black in white light. Substances which absorb radiant energy outside the limits of the visible spectrum, although appearing white to the eye, show characteristic absorption 'fingerprints' when analysed with a spectrophotometer.

It was pointed out by Bury in 1935 that many organic dyestuffs could be regarded as resonance hybrids, that is existing in a low energy form intermediate between the possible alternatives—which could be produced in theory by rearranging the valencies within the molecule (canonical forms). Thus, in the case of the indicator methyl orange in acid solution, a resonance hybrid intermediate in structure between the following two canonical forms was thought to exist.

$(CH_3)_2\overset{+}{N}=\underset{}{\bigcirc}=N-\underset{H}{N}-\underset{}{\bigcirc}-SO_3H$

$\Updownarrow$

$(CH_3)_2N-\underset{}{\bigcirc}-N=\underset{+}{\underset{H}{N}}-\underset{}{\bigcirc}-SO_3H$

canonical forms of methyl orange in acid solution

It was difficult to see why resonance hybrid structures of this kind should produce absorption of light in the visible range and hence appear coloured, until the application of quantum and wave mechanics to the problem. This revealed that resonance itself was not the fundamental factor responsible for producing colour, but that structures of this type were much more likely to absorb radiation within the visible part of the spectrum.

The absorption of a small 'packet' of light energy (photon) by a molecular structure can only occur if the energy contained in the photon is exactly the amount required to promote one of the electrons in the molecule to a molecular orbital of higher energy than the one it normally occupies (ground state). The photons associated with visible light are of comparatively low energy and are only absorbed, therefore, by structures which have small energy differences between their molecular orbitals. Resonance hybrids possess less energy (resonance energy) than their corresponding canonical forms and are therefore more likely to be able to absorb the low energy photons corresponding to the wave-lengths of visible light.

The intensity of energy absorption by a molecule depends upon the tightness with which the electron orbitals are coupled to the system. If the coupling is too tight then the electrons are not free to resonate (forbidden transition) and any incident energy is converted into heat before the electron can absorb sufficient of it to migrate to a more energetic orbital. A simple analogy to this is a piano string which cannot resonate with other vibrating strings if it is being coupled tightly to the piano frame by the damper.

The electrons involved in forming single bonds between atoms (σ- or sigma-electrons) are tightly coupled and absorb in the far ultraviolet region. In multiple bonds, however, another type of electron (π- or pi-electron) is found which is more loosely coupled and so has a smaller energy difference between its orbitals. The π-electrons are therefore able to absorb in the ultraviolet and visible regions. Thus, to produce a dye it is necessary to increase the π-electron character of a molecule by introducing double-bonded structures. The introduction of nitrogen atoms is particularly effective in this respect.

The conjugation of double bonds in a molecule lessens the coupling of the electrons still further and also lowers the associated energy levels. This promotes absorption of the lower frequency radiation at the red end of the visible spectrum and hence a shift of the colour of the compound towards the blue end of the spectrum (bathochromic effect). This effect is a useful indication of the extent to which a molecular structure is conjugated.

The unique conjugated nature of the benzene ring explains its presence in virtually all organic dyes and the bathochromic effect produced by increasing the range of conjugation is well demonstrated by considering the following series of condensed aromatic rings:

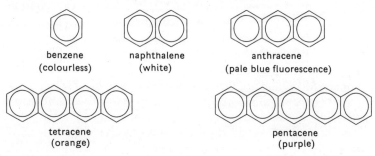

benzene
(colourless)

naphthalene
(white)

anthracene
(pale blue fluorescence)

tetracene
(orange)

pentacene
(purple)

Another conjugated double-bond system frequently found in dye molecules is the quinonoid structure.

The range of conjugation can be extended by forking or cross-conjugation, thus converting a single conjugated chain into a double one. A good example of this occurs in the unsaturated cyclic hydrocarbon fulvene, which is yellow, whereas the corresponding straight chain alkene, hexatriene, absorbs in the ultraviolet region.

conjugated hydrocarbon chain
(hexatriene)
no absorption in visible spectrum,
colourless

cross-conjugated hydrocarbon chain
(fulvene)
absorption in visible spectrum,
yellow colour

The deep blue colour of indigo is also due to cross-conjugation of this type.

cross-conjugated chain
(indigo—deep blue)

Deepening of colour can also be due to the lengthening of a conjugated chain by hyper-conjugation. This effect is commonly seen when the hydrogen atoms of an amino group are replaced by methyl groups (methylation). Although the single bonds associated with the methyl groups only contain σ-electrons they behave in some ways like π-bonds and conjugate to a limited extent with the π-electrons of the nitrogen atom, thus extending the conjugation chain.

The methylation of the triphenylmethane dye magenta to give the deeply coloured methyl violet has already been mentioned (see formula on page 195) and is a good example of hyper-conjugation.

If the conjugation bridge between two chromophoric groups is broken by the inclusion of a length of saturated hydrocarbon chain, the latter acts as a kind of insulating barrier to the movement of π-electrons which normally occurs along a conjugated chain. The light absorption of the two chromophore units takes place independently under these circumstances, giving an additive effect. This is often made use of in increasing the shade range of azo dyes. Thus a green dye can be produced by synthesizing a molecule containing a blue and a yellow chromophore linked by a saturated hydrocarbon chain.

It has recently been shown that the plane of incident light also has an effect upon its absorption by a dye molecule. Maximum absorption occurs when the plane of the radiation and that of the molecular orbitals of the dye molecule coincide. When the molecule has two orbital planes, light of two different wavelengths can be absorbed according to the plane of the incident light. The fluorescein molecule behaves in this way, solutions appearing red by transmitted light and green by reflected light (dichroism).

Thus it will be seen that the prime requirement for a coloured molecule is a marked π-electron character. This can be produced by the inclusion of aromatic rings and other conjugated hydrocarbon systems which usually contain nitrogen atoms, and occasionally other 'hetero' atoms such as oxygen and sulphur. This enables the region of absorption to shift from the ultraviolet into the visible spectrum which in turn results in the reflection of the appropriate complementary colour. Certain groups (Witt's auxochromes) such as —OH, —NH$_2$, and —OCH$_3$ provide lone electron pairs which are

212   *Fuels, Explosives and Dyestuffs*

able to conjugate with the π-electrons of the chromogen double bonds. This causes a further shift in absorption towards the red end of the spectrum, thus deepening the colour.

Optical dyes or 'fluorescers' have a similar conjugated structure to dyes and absorb strongly in the near ultraviolet region. The nitrogen atoms present in such molecules are, however, firmly held in rigid ring structures. This results in such weak coupling of the π-electron orbitals that most of the radiant energy absorbed is re-emitted as visible light instead of being dissipated throughout the molecule. Such materials are often incorporated in soap powders to produce fluorescence of white laundered articles in daylight.

an example of a 'fluorescing' structure

THE CLASSIFICATION OF DYES

Dyestuffs are widely used for colouring such diverse materials as paper, plastics, leather, wax polish, anodized aluminium and cosmetics. A few carefully selected dyes such as cochineal are also permitted as colorants in the manufacture of sweets, table jellies, custard powder, ice-cream and other foodstuffs.

By far the most important application of dyes, however, is to colour textile fibres. For this purpose they need to have a number of special properties. They must be reasonably cheap, applicable in solution or as a dispersion, and be as fast as possible to light and laundering. In addition they must be capable of withstanding the finishing processes carried out on the textile during manufacture.

Dyes can of course be classified according to their chemical nature, but it is usually more convenient to consider them according to their method of application and the type of fibre to which they are most suited. This is usually reflected in the trade name given to a dye by the manufacturer. Thus ICI use the names 'Caledon' for vat dyes, 'Lissamine' for acid wool dyes, 'Chlorazol' for direct dyes, 'Procion' for reactive dyes and so on. Other symbols are often added to this type-name to indicate fastness, colour shade and other information. For example 'Orange G' and 'Orange R' signify a yellowish orange and a reddish orange respectively, similarly 'Red G' is a scarlet and 'Red B' a bluish red. The trade name 'Chlorazol Fast Orange G' will therefore inform the dyer that this is a yellowish orange direct dye which is fast to light.

Until recently, however, difficulties arose because different manufacturers used their own notations to describe dyes and this often resulted in the same dyestuff being marketed under a variety of names. To avoid this confusion the Society of Dyers and Colorists of Great Britain instituted a colour index (C.I.) in which every dye was given a three-part systematic name. The first part of the C.I. nomenclature refers to the dye type, the second gives its colour and the third is a serial number. Thus 'Chlorazol Fast Orange G' mentioned above is listed as 'C.I. Direct Orange 23'.

DYE TYPES

(1) MORDANT DYES

The natural colouring matters such as logweed, cochineal and madder which were used in the early days of dyeing did not have a great affinity for textile fibres. Materials such as cotton were therefore treated before dyeing to make them more receptive by impregnation with metallic salts known as mordants. This word was derived from the French *mordre* meaning to bite, as it was considered that the mordant, after attaching itself to the fibre, was able to bite into the dye and hold it in place.

Dyes of this kind are termed mordant dyes, the most common mordants being soluble salts of aluminium, chromium, tin, iron and copper. The fabric to be dyed is steeped in a solution of an appropriate salt such as aluminium acetate and then steamed. This converts the salt into the corresponding metal hydroxide which, being bulky and insoluble, locks into the fibre structure. On treating the mordanted fibre with a dyestuff in solution, an insoluble, stable, coloured complex is formed between the dye molecules and the metallic hydroxide which is called a 'lake'.

The metal atoms of the mordant are electron acceptors and are able therefore to accept electrons from donor groups such as hydroxyl (—OH), carbonyl ($\text{>}C=O$) and diazo (—N=N—), which are commonly found in dyestuff molecules. Alizarin, for example, with a pair of both carbonyl and hydroxyl groups readily forms a red complex with aluminium.

alizarin

alizarin complex with aluminium

Many mordant dyes give different shades of colour with different mordants and these are said to be 'polygenetic'. Mixed mordants are sometimes used to obtain a variety of shades with polygenetic dyes. Alizarin, one of the oldest mordant dyes, in addition to the red complex formed with aluminium mordants, gives a pink colour when mordanted with tin, and a range of browns with iron, chromium and copper mordants. Alizarin, extracted at one time from madder grown in Turkey, gave exceptionally bright reds with alum mordants, which led to the use of the term Turkey Red, the wetting agent used during the dyeing process being known as Turkey Red Oil.

Cochineal, which is extracted from the female of a small species of Mexican insect (*Coccus cacti*) found on the prickly pear, contains a mordant dye called carminic acid and gives a bright scarlet when mordanted with tin, which has been extensively used in the dyeing of ceremonial uniforms and hunting coats.

In 1889 Nietzki discovered that if a salicylic acid unit containing adjacent carboxyl and hydroxyl radicals could be introduced into a dye molecule, it was possible to introduce a chromium mordant after dyeing ('afterchrome' mordanting). Five years later Sandmeyer showed that afterchroming could be carried out even more readily using dihydroxy azo dyes such as 'Eriochrome Black T' (C.I. Mordant Black 11). This process is specially suited to dyeing wool and gives a very fast colour. Because of this it is used for dyeing articles like men's woollen socks which are washed frequently.

Eriochrome Black T

The wool is dyed initially in a bath acidified with acetic acid. When the dyebath is exhausted, potassium or sodium dichromate is added and the cloth boiled for 30–40 minutes. It is essential to use up all the dye before chroming and 'exhausters' commonly used are sulphuric or formic acid. In an alternative 'metachrome' or 'synchromate' process the chroming and dyeing are carried out simultaneously in a dyebath containing sodium dichromate, ammonium sulphate and sodium sulphate. This has obvious advantages in saving of time, labour and plant, besides enabling easier colour matching. Occasionally chrome mordanting is carried out by boiling the wool before dyeing in a sodium dichromate solution which has either been acidified with sulphuric acid or reduced with potassium hydrogen tartrate (cream of tartar).

The success of the chrome dyeing processes sparked off research into the possibility of producing soluble chrome lakes from azo mordant dyes so that they could be applied directly to wool. This resulted in the production of the 'Neolan' range of dyes by Ciba Ltd in 1915, followed by the 'Palatine' dyes. These are known as premetallized dyes, and are applied at the boil in a bath containing ammonium acetate.

(2) ACID DYES

These are important protein dyes and are widely used for wool dyeing. They owe their name to the fact that they are carboxylic or sulphonic acid derivatives and are used in an acid dyebath. Most of the acid dyes are available as soluble sodium salts, the free dye acids being hygroscopic and therefore difficult to store and handle. The strong affinity of acid dyes for wool and silk is due to the salt formation which occurs between the active coloured anion of the dye and the amino groups ($-NH_2$) of the protein. A similar reaction occurs with polyamides such as nylon, and certain acrylic fibres. Cotton on the other hand does not normally dye satisfactorily with an acid dye unless specially treated to incorporate a basic component ('Rayolanda' process).

This group of dyes includes a number of different chemical types including azo, triphenylmethane, and anthraquinone derivatives. The first acid dyes, however, were produced by sulphonating basic dyes such as aniline (Bleu de Lyons—1862) and magenta (Acid Magenta). It was not until 1876 that the first azo acid dye was prepared by coupling β-naphthol and diazotized sulphanilic acid (Orange II).

Orange II (C.I. Acid Orange 7)

Two other pioneer azo acid dyes were Azo Geranine and Tartrazine. A little later in 1890 the first acid dyes derived from anthraquinone appeared and proved to be exceptionally fast to washing. These were applied in a neutral bath and termed acid milling dyes because they were originally used for dyeing wool

Dyestuffs

which was to be milled or felted. Modern varieties include the 'Coomassie' range manufactured by ICI, some of which have a marked affinity for cellulosic fibres, as, for example, 'Coomassie Brilliant Blue R' (C.I. Acid Blue 83).

The triphenylmethane type acid dyes, including the ICI range of 'Lissamine' dyes, were pioneered in 1862 by Nicholson with the discovery of 'Xylene Blue VS'.

Xylene Blue VS (C.I. Acid Blue I)

These require a strongly acid dyebath and, because of their good dyeing properties, are known as level-dyeing acid dyes. They are light fast but their fastness to washing is markedly inferior to the acid milling dyes. Level-dyeing acid dyes give more even dyeing if sodium sulphate is added to the dyebath but this 'assistant' is only effective if the bath is strongly acidic (below pH 4·7). Therefore sodium sulphate cannot be used as a levelling agent for acid dyes which are used in weakly acid or neutral baths.

Picric acid (2,4,6-trinitrophenol) is an interesting example of an acid dye because it does not possess a carboxylic or sulphonic acid grouping. The three electron-withdrawing nitro groups cause such a powerful shift of electrons away from the hydroxyl hydrogen that the latter readily ionizes producing a strongly acid substance.

(3) BASIC DYES

Many of the early dyes such as Magenta, Crystal Violet, Fuchsine and Mauveine belong to this group. Unlike the acid dyes the cation is in this case the coloured component. The dye base is colourless and has the general structure shown below. On forming a salt such as the hydrochloride, oxalate or a double salt with zinc chloride, a brilliant colour is produced.

$$HOR_1-\underset{R_2}{\underset{|}{\bigcirc}}-N\overset{R_2}{\underset{R_2}{\diagdown}} \xrightarrow{\text{dil. HCl}} \left[R_1=\bigcirc=\overset{+}{N}\overset{R_2}{\underset{R_2}{\diagdown}}\right]Cl^- + H_2O$$

generalized formula of 'colour base' of basic dye   coloured salt of basic dye

(where R_1 = an aromatic radical; R_2 = usually —CH_3 or —H)

The unusually intense hues of the basic dyes are unfortunately not fast to light and only moderately fast to washing. Neither are they easily soluble in water although they readily dissolve in alcohol. Most of the basic dyes are sensitive to heat (thermo-labile) and partially decompose on boiling. They have no affinity for unmordanted cellulosic fibres but are able to form salt links with the free carboxyl groups of protein fibres. So great is the affinity of basic dyes for protein fibres that usually a retarding agent such as acetic acid is used in the early stages of dyeing to control the rate of absorption. The powdered dye is made into a paste with 30% acetic acid and then stirred into hot water, dyeing being carried out at 80°C, or slightly lower in the case of very thermo-labile dyes such as Auramine.

The fastness of basic dyes can be improved by after-treatment with a tannic acid solution followed by agitation in a bath of potassium antimony tartrate (tartar emetic). This produces an insoluble antimonyl/tannic acid complex with the dye.

A similar process is used to treat cellulosic fibres which possess no acidic groups and therefore cannot form salt links unless mordanted. The tedious tannic acid/tartar emetic treatment is avoided by using synthetic mordants such as Taninol BMN and Katanol which have a direct affinity for cellulose and are insensitive to traces of metals such as iron, which affect tannic acid. The cellulosic material is immersed in very hot water (90°C) containing the mordant together with a wetting agent such as Turkey Red Oil (sulphonated castor oil), for 1–2 hours before dyeing at about 40°C.

Although the dyeing of cellulosics with basic dyes has declined,

Dyestuffs

new interest has been kindled in the affinity of these dyes for acrylic fibres such as 'Dynel' and 'Orlon'. The fastness to both light and washing of basic dyes on this type of fibre is unaccountably high.

Most of the well-known basic dyes, such as Magenta, Crystal Violet, Malachite Green and Para Rosaniline, are triphenylmethane derivatives. Other members of this group, however, include azo derivatives (Bismarck Brown), azines (Neutral Red), thiazines (Methylene Blue) and xanthenes (Rhodamine Red).

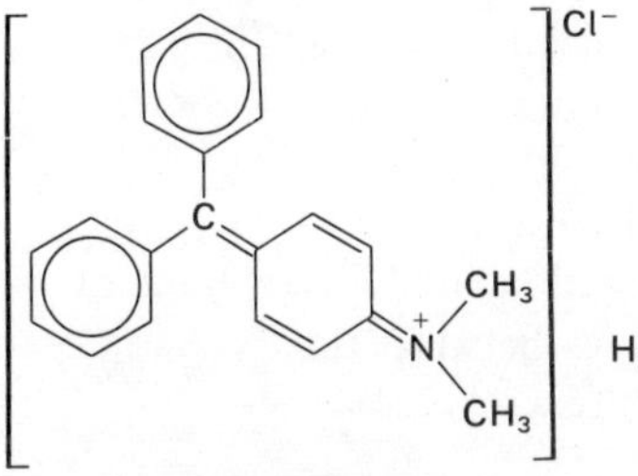

Malachite Green

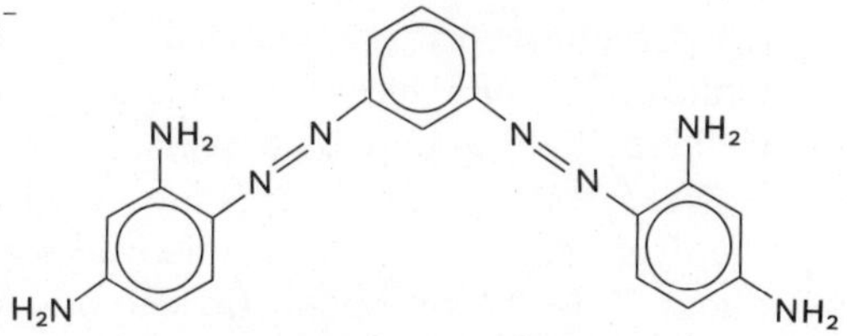

Bismarck Brown

(4) DIRECT DYES (SUBSTANTIVE DYES)

These are the most important dyes for cellulosic textiles and as their name suggests they can be applied without mordanting. They also have an affinity for protein fibres and are therefore useful for dyeing wool and cotton mixtures. Despite their indifferent fastness to washing and light the direct dyes have always been popular because of their cheapness and ease of application.

Most of the direct dyes are the sodium salts of sulphonated azo compounds in common with the earliest member of the family, Congo Red, discovered by Böttiger in 1884. The molecules of this group of dyestuffs are often bulky, some containing multiple azo groupings, such as the triazo dye 'Diazo Brown 3 RNA CF', and the tetrakis azo dye 'Chlorazol Brown GM'.

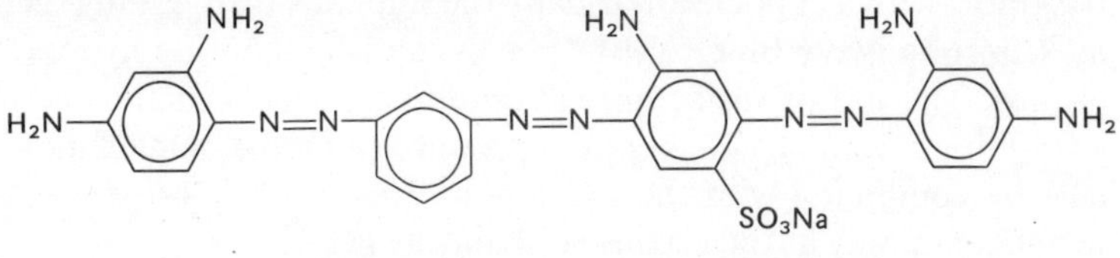

Diazo Brown 3RNA CF (C.I. Direct Brown 138)

Fuels, Explosives and Dyestuffs

Others have complex structures built around a centrally coordinated metal atom, as in the case of the phthalocyanins.

Direct dyes are usually applied from a hot aqueous solution to which salt is added to decrease the solubility of the dye and produce level dyeing. To improve wet-fastness direct dyes are often used which contain a primary amine grouping. This group can be diazotized after dyeing and then developed using coupling agents such as β-naphthol, resorcinol, m-phenylenediamine and aminodiphenylamine ('Fast Blue Developer P') producing an insoluble dye of high wet fastness. Direct dyes of this kind include 'Chlorazol Orange TR' and 'Diazamine Blue G', although of course the shade is altered by the coupling process.

Wet-fastness of direct dyes has also been improved by steeping the dyed fabric in 3% formaldehyde solution for half an hour at about 75°C. This forms methylene bridges between the dye molecules and locks them into the fibre structure.

$$2 \quad \underset{\text{direct dye}}{R-N=N-\underset{H_2N}{\bigcirc}-NH_2} \quad \xrightarrow[75°C]{3\% \; H \cdot CHO}$$

$$\underset{\text{dye molecules linked by methylene bridge}}{R-N=N-\underset{H_2N}{\bigcirc}\underset{NH_2}{\overset{CH_2}{-}}\underset{H_2N}{\bigcirc}-N=N-R} \; + \; H_2O$$

(where R = polycyclic structure)

Fastness to both light and washing can also be achieved by after-treatment with a copper salt, usually the sulphate, on dyestuffs such as 'Cuprofix Navy Blue CGBL' possessing neighbouring hydroxyl groups. The use of surfactant cationic fixing agents such as cetyl pyridinium bromide ('Fixanol C') produces a more complex molecule by combining with the dye anion. This again improves wet-fastness but only at the expense of light-fastness.

(5) VAT DYES

The vat dyes are of particular interest both because of their historical significance and their unique method of application. The woad of the ancient Briton, and the Tyrian Purple of the Roman Emperor, were both dyes of this kind. Indigo also has a long history, at one time being extracted from a plant (*Indigofera tinctoria*) originally cultivated in India (Gk. *indikon*—Indian (dye)) then spreading to the Middle East and Europe. Edward III even encouraged the growing of the indigo plant in Britain but this was later abandoned owing to the opposition of the woad growers.

The death blow to the production of vat dyes from vegetable sources was the commercial synthesis of indigo by Badische Anilin und Soda-Fabrik at the turn of the century. Aniline was first converted into phenylglycine which was then fused with caustic soda at 350°C to give the sodium salt of indoxyl. This was then oxidized by an air blast and converted into indigotin.

(a) aniline $\xrightarrow[\text{heat}]{CH_2ClCOOH}$ phenylglycine $\xrightarrow[350°C]{NaOH}$ indoxyl

(b) 2 indoxyl $\xrightarrow[\text{(air blast)}]{\text{atmospheric oxidation}}$ indigotin

A more recent process employed by the Germans in World War II uses aniline and ethylene oxide as starting materials.

(a) [reaction scheme: aniline + ethylene oxide → N-(2-hydroxyethyl)aniline → (NaOH/NaNH₂, 200°C) → sodium indoxylate]

(b) [reaction scheme: 2 × sodium indoxylate → (1. hydrolysis, 2. air blast) → indigotin]

indigotin

Like the other vat dyes, indigotin is insoluble in water and must be converted into a soluble form before being used for dyeing. Originally this was done by a fermentation process, a suspension of the dye being warmed in a vat with bran and lime. After several days the dye was reduced to a soluble 'leuco' compound called indigo white (Gk. *leukos*—white) and formed a watery green solution. On soaking the material to be dyed in this solution and then exposing it to the air, the indigo white re-oxidized to the blue insoluble form within the fibres, thus producing a very fast finish.

The fermentation process was tedious and difficult to control and chemical processes were eventually worked out which superseded it. One of the early techniques for solubilizing indigotin made use of freshly precipitated ferrous hydroxide prepared in the dye vat by adding lime water to ferrous sulphate solution. Later a solution of zinc in lime water (calcium zincate) was used in the same way, but neither of these processes was wholly satisfactory as the bulky precipitates of the hydroxide and zincate interfered with the uptake of the dye by the fibres.

Much more successful was the use of an alkaline solution of sodium hydrosulphite ('dithionite'). The formation of the indigo white anion was the result of electron addition rather than hydrogenation, the hydrosulphite ion ($S_2O_4^{--}$) acting as an electron donor.

Impregnation of fabrics with the leuco dye is often carried out at room temperature (20°C), although for some vat dyes tempera-

indigotin → (Na$_2$S$_2$O$_4$ + NaOH) → anion of indigo white

tures up to 58°C are necessary. The degree of alkalinity required for vat dyeing also depends upon the nature of the dye. Indigoid dyes require only weak alkalis such as sodium carbonate or ammonia, the anthraquinone dyes on the other hand are used with a caustic alkali such as sodium hydroxide. After absorption of the alkali leuco dye the fabric is squeezed to remove excess liquor and then oxidized. Almost all the leuco compounds of vat dyes can be oxidized by exposure to the air for half an hour or so. To speed up this process or to oxidize material in rolls, or as yarn on bobbins ('cheeses' or 'cones') it is necessary to use an oxidizing solution. Originally acidified potassium dichromate or hydrogen peroxide solutions were used, but more recently solutions of sodium perborate or percarbonate have become popular. Any excess indigotin deposited upon the fibres is removed by final treatment in a hot soap bath which also serves to aggregate the dye particles within the fibres.

Where the dye has a low affinity for the fibre to be dyed, as in the case of indigotin on cotton, the desired shade is built up by repeated dipping and oxidation. The dyeing of protein fibres with vat dyes is limited because of the danger of the alkaline dyeing solutions causing damage to the fibre structure.

Fibre damage to cellulosic materials occurs when they are dyed with certain orange or yellow vat dyes and then exposed to sunlight for lengthy periods. No satisfactory solution has yet been put forward for this photochemical degradation although it has been found that vat blues fade more rapidly in combination with vat yellows.

A big step forward in the history of vat dyes was the synthesis in 1921 of a soluble form of indigotin known as Indigosol O. This was the sodium salt of a sulphate ester of indigo white prepared by treating a solution of the latter in pyridine with chlorosulphonic acid and then neutralizing with sodium hydroxide solution.

$$\text{indigo white} \xrightarrow[\text{(in pyridine)}]{\text{ClSO}_2\text{OH}} \text{sulphate ester of indigo white (Indigosol O = sodium salt)}$$

Indigosol O was not only water soluble but stable in air and had a marked affinity for cellulose fibres. Soon, other anthraquinone and indigoid type soluble vat dyes were produced, all having similar properties to Indigosol O. Dyes of this kind are applied from an aqueous solution containing a mild oxidizing agent such as acidified sodium nitrite. Development of the colour takes about 15 minutes and is followed by boiling in a soap bath.

The 'Soledon' range of soluble anthraquinone vat dyes were discovered in 1924 by Scottish Dyes Ltd (now a subsidiary of ICI Ltd). These are prepared directly from the vat dye itself and not the leuco form, a solution of the dye in pyridine being treated with chlorosulphonic acid in the presence of metallic copper or iron. 'Soledon' dyes are still extensively used for dyeing cellulosic fibres.

(6) SULPHUR DYES

These are prepared from organic compounds, such as phenols and phenylamines, by heating them with sulphur or sodium polysulphide solution. They are unique as dyestuffs in that they all possess sulphur linkages in their molecular framework. Like the vat dyes they have to be reduced to render them water soluble, the reducing agent used being a slightly alkaline solution of sodium sulphide. It is thought that the sodium sulphide attacks the sulphur links of the dye, thus splitting it up into a simpler leuco form which will dissolve in water and is attracted to cellulose fibres. The leuco form is readily oxidized back to the insoluble form, the whole process being very similar to that used for vat dyeing.

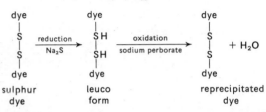

The sulphur dyes are very fast to washing but have rather dull, flat colours and have not achieved great popularity, although they are cheap and still widely used for dyeing cotton and viscose, and to a limited extent, nylon.

A brown sulphur dye was prepared as early as 1873 by heating sawdust with sulphur and sodium hydroxide. The first commercially important dyes of this type, however, were the Sulphur Blacks discovered by Vidal in 1893 and prepared from p-aminophenol or p-phenylenediamine by fusion with sodium sulphide and sulphur. Later blue, orange, green, brown and yellow dyes were produced. Stainless steel vats are commonly used in the dyeing process because of the corrosive nature of the hydrogen sulphide liberated by solutions of sulphur dyes and alkaline sodium sulphide.

Solubilized sulphur dyes have been marketed by ICI in the 'Thional' range and are distinguished by the suffix M, such as 'Thional Sky Blue 6BM'. They are prepared from the traditional sulphur dyes as a stable leuco form which is easier to use, although the addition of sodium sulphide is still required when dyeing cellulose fabrics.

(7) AZOIC DYES

The water soluble direct and acid dyes have already been mentioned. The azo dyes of this group however are all insoluble in water and are produced within the fibres themselves. This is achieved by impregnating the fibres with a naphthol or phenol and then coupling with a diazonium salt—a process known as ingrain dyeing. The first successfully developed azoic dye was Para Red, produced by coupling β-naphthol with diazotized p-nitroaniline.

The cheapness of this novel type of dye, coupled with its extreme fastness to washing, made it commercially attractive for dyeing

Fuels, Explosives and Dyestuffs

β-naphthol (2-naphthol) + p-nitroaniline diazonium chloride → (in ice water, coupling) → Para Red (deposited in textile fibre)

cotton, and up to 1922 a number of azoic dyes were produced by coupling β-naphthol with a variety of diazonium salts. In that year β-naphthol was replaced as a coupling agent by derivatives of β-hydroxynaphthoic acid, such as the anilide which was prepared by ICI and called 'Brenthol AS'. A similar range of β-hydroxynaphthoic acid derivatives to the 'Brenthols' were produced in Germany by I.G. Farbenindustrie as 'Naphtols'. The great advantage of these new coupling components was that, unlike β-naphthol, they had an affinity for cellulose fibres. Also a wide range of shades could be obtained by ringing the changes on the large number of coupling agents available. The Colour Index now contains the staggering total of 52 azoic coupling agents and 50 bases.

One of the drawbacks to the use of azoic dyes in the early days

naphthol AS. AN (C.I. Azoic Coupling Agent 27) + m-nitroaniline diazonium chloride → (in ice water, coupling) → orange azoic dye

Dyestuffs

was the inconvenience of preparing diazonium salts. These were chemically unstable even at low temperatures and their bottle life unpredictable. There was a demand therefore for a stable form of diazonium salt which could be stored indefinitely and used directly in the dyebath without preliminary preparation. Research into this problem resulted in the development of the 'Fast Salts', such as 'Fast Scarlet Salt LG' and 'Fast Blue RR Base', which could be dissolved in water when required and were then ready for coupling. The coupling rate of these salts is related to the pH of the dyebath used—the most active diazonium salts requiring a strongly acid bath (pH 4–5), those with weak coupling energies requiring an alkaline bath (pH 7–8·2).

The early stabilized diazonium compounds were known as antidiazotates and regenerated the active diazonium salt on acidification. More recently 'Rapidogen' diazo compounds have been produced for fabric printing. These contain diazo-amino compounds possessing water-solubilizing groups and can be mixed with a naphthol-coupling agent and printed on to fabric. A coupled dyestuff can then be produced by steaming or acidifying the treated material. A commonly used diazo-amino compound is produced by reacting 2,5-dimethoxyphenyl diazonium chloride with the sodium salt of sarcosine (methylglycine)—an amino acid found in muscle tissue.

(8) OXIDATION DYES

This is a small and relatively unimportant group of dyes produced by the oxidation of aniline hydrochloride (Aniline Black) and related aromatic amines. Oxidation dyes are used on cotton and viscose, the dyeing process being carried out in two stages. The material to be dyed is first impregnated with a solution of the amine salt containing a catalyst (copper chloride) and an oxidant (sodium chlorate). After steaming, the oxidation process is completed by running the fabric through an oxidizing bath containing hot sodium dichromate solution. Care has to be taken not to damage the cellulose fibres during oxidation and recently *p*-amino diphenylamine has been used instead of aniline, since this can be used under milder oxidizing conditions.

(9) DISPERSE DYES

Cellulose acetate fibres are hydrophobic and attempts to dye them using aqueous solutions of basic dyes were not satisfactory. It had been noticed, however, that acetate fibres readily absorbed insoluble dyestuffs when present in the dyebath in the form of a fine suspension. Ellis and Baddiley in 1924 developed a method of dispersing finely ground insoluble azoic pigments and amino derivatives of anthraquinone using sulphonated ricinoleic acid (castor oil) as a dispersing agent. As these SRA dyestuffs were simple to apply and produced a satisfactory fast finish a number of manufacturers became interested in their production. Within a few years a variety of disperse dyes appeared on the market, such as the 'Dispersols' (ICI), 'Cibacets' (Ciba), 'Cellitons' (I.G. Farben) and 'Setacyls' (Geigy).

The absorption of dye particles by cellulose acetate fibres only takes place satisfactorily if the particle size is extremely small,

Dispersol fast yellow G
(C.I. Disperse Yellow 3)

Cibacet orange 2R
(C.I. Disperse Orange 3)

thus enabling a large area of surface contact. Meyer and others (1926) showed that the pigment probably formed a solid solution in the acetate fibres in a manner analogous to the distribution of a solute between two immiscible solvents. More recently, however, this has been shown to be an over-simplification and it is thought that the dye is absorbed in the molecular state and held in position by hydrogen bonds and Van der Waals forces.

Disperse dyes containing amino ($-NH_2$) groups can be diazotized and coupled after absorption to produce a further range of useful shades, especially for deep colours such as navy blue and black. Partially solubilized disperse dyes have been produced by incorporating $-OSO_3H$ and $-COOH$ groups in the molecule but this considerably reduces the uptake of the dye on acetate fibres and is little used.

Special difficulties are encountered in the disperse dyeing of cellulose triacetate ('Tricel') since all the hydroxyl units of the cellulose chain have been acetylated. Thus the resulting fibres are even more water-repellent than those of secondary cellulose acetate. To overcome this difficulty the temperature of the dyebath is raised from 80°C (the temperature used for secondary acetate) to boiling. In addition swelling agents such as acetone, acetic acid and diethylene glycol diacetate (DEGDA) have been used to pre-treat triacetate fibres to make them more receptive.

Using a pressurized dyebath at a temperature above 100°C, and a dye-carrier such as o-phenylphenol, 'Terylene' and 'Rhovyl' synthetic fibres have also been successfully coloured with disperse dyes.

Two disadvantages of disperse dyes are a tendency to sublime at high temperatures, such as would be encountered during ironing, and a peculiar discoloration known as gas-fading, which is prevalent in the atmosphere of large towns. Gas-fading is particularly serious with blue disperse dyes based upon anthraquinone, which turn first red and then grey. The cause of gas-fading has been traced to the presence in the atmosphere of small amounts of nitrous oxide produced by burning gas flames and the interaction of atmospheric oxygen and nitrogen on red-hot surfaces such as electric fire elements. A certain amount of protection against gas-fading is obtained by using anti-gas-fading agents such as zinc oxide, triethanolamine, diphenyl ethylene diamine and 'Tumescal D' (ICI).

(10) REACTIVE DYES

One of the major problems in producing a dyestuff is that it must be soluble enough to be easily applied in solution and yet once it is deposited on the fibre it must be so insoluble as to be resistant to frequent washing. The types of dyestuffs mentioned above reveal a number of ingenious ways of resolving this difficult problem by mordanting, ingrain and developed dyeing, vat dyeing and so on. At the end of World War II, however, a fundamentally different approach was made to the problem.

The attention of research workers was turned towards the properties of dyestuffs containing cyanuric chloride residues (triazinyl dyes).

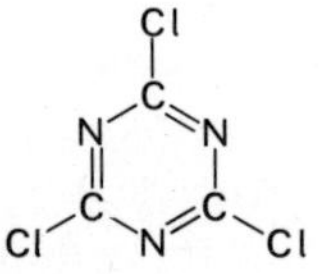

cyanuric chloride

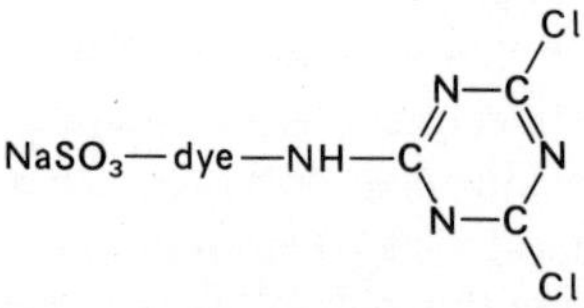

triazinyl dye

These were able to couple directly with the hydroxyl groups of the cellulose molecule in the presence of alkali and also the amino groups present in the polypeptide chains of wool and silk, and the imino groups (—NH—) of nylon polymers.

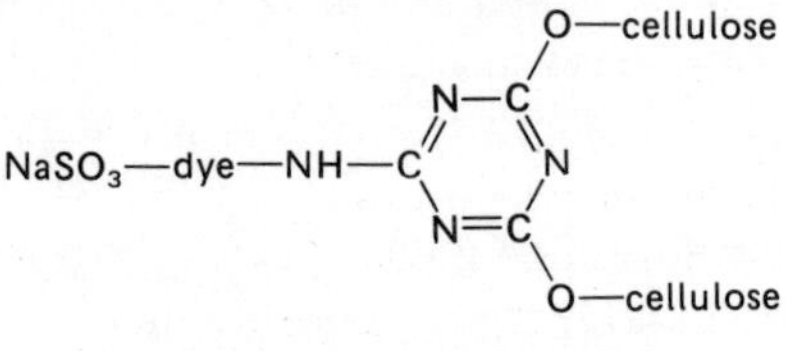

cellulose attached to a triazinyl dye residue by means of ether linkages

It will be seen that the dye molecule chemically combines with the fibre to form a stable compound which is unaffected by washing and is usually also light-fast. Evidence of the formation of a new compound is given by the ineffectiveness of the usual reagents used for stripping dyes from cellulose fibres and their insolubility in cuprammonium solutions. The first dyes of this type were the

Dyestuffs

'Procion' dyes produced by ICI in 1956 as the result of earlier work by two of their research chemists Rattee and Stephen. In the following year the Swiss firm of Ciba produced a similar range of 'Cibacron' dyes which were soon followed by the 'Remazols' (Hoechst), 'Reactones' (Geigy) and 'Drimarenes' (Sandoz). The Remazols have a structure as shown below, alkaline hydrolysis producing a reactive vinyl grouping which reacts with the —OH group of cellulose as before.

$$\text{dye} - SO_2 - CH_2 - CH_2 - O - SO_3H \xrightarrow[\text{hydrolysis}]{\text{alkaline}} \text{dye} - SO_2 - CH = CH_2$$

$$\text{dye} - SO_2 - CH_2 - CH_2 - Cl \qquad \text{active vinyl sulphone}$$

alternative general forms of 'Remazol' dyes

The reactive dyes are readily soluble in water and reaction with cellulose or protein fibres is fairly rapid. Mixtures of reactive dyes can also be used to produce a range of shades as they do not interfere with each other in solution.

Normally the goods to be dyed are run for about 10 minutes in a solution of the dye at 20–30°C and then salt is added and dyeing continued for a further 15 minutes or so until the required shade has been reached. The dye is then fixed by adding alkali (sodium carbonate) and running for a further hour or more. Solutions of active dyes do not keep well because hydrolysis occurs with the production of non-reactive water soluble dyes which have an affinity for cellulose.

triazinyl dye $\xrightarrow{H_2O}$ water soluble non-reactive dye

Fortunately these non-reactive dyes are not fast and can easily be removed from the dyed material by washing. Hydrolysis is nevertheless reduced to a minimum because of the loss of dye involved.

Recently a 'Procion'-resin process has been used in which the reactive dye is mixed with a partly cured resin. After impregnation with a solution of this mixture containing a wetting agent and lubricant (Velan NW), heat-curing is carried out producing a simultaneous dyeing and resin finishing. Other recent advances have been the development of a range of reactive dyes containing a pyrimidine residue instead of the triazine nucleus and the introduction of reactive disperse dyes by ICI in 1959 ('Procinyl') dyes.

DYEING SYNTHETIC FIBRES

A textile fibre consists of very large numbers of long polymer molecules orientated together into strands and containing between them minute capillary channels. It is possible to measure the average diameter of these capillaries ('mean pore size') and considerable differences in pore size have been found in different fibres. Man-made fibres, especially the synthetics, have far smaller pores than natural fibres.

The pore size of a fibre also varies in size according to the temperature and the nature of the medium in which it is immersed. When a fibre is immersed in water, or other liquid containing molecules small enough to penetrate its pores, it swells. Pore shrinkage occurs on the other hand when a fibre is immersed in a strong salt solution. The size of dye particles in suspension can be affected in a similar way by temperature and the presence of electrolytes.

The dyeing process depends basically upon two factors:

(1) The penetration of the tiny channels between the mass of fibrous polymer by particles of dye—rather as ink spreads along the pores of blotting paper.
(2) The fixing of the particles within the fibre pores after penetration to make the colour fast to washing.

The dyeing of synthetic fibres presents special difficulties. The high degree of orientation of cold-drawn fibres, such as nylon and 'Terylene', which gives them such high strength (tenacity) also results in a very small pore size which is often inaccessible to even small molecules like water. In addition to this the synthetic fibres, containing long paraffin-type chains, are water-hating (hydrophobic). The effect of this is clearly seen in comparing the uptake of water by different fibres (imbibition).

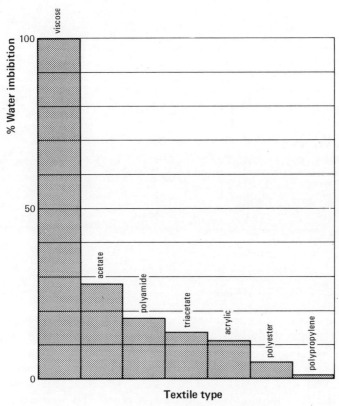

Fig. 5.1 Uptake of water by textile fibres

It is interesting to review the dyes and techniques which have been used to meet the challenging problems involved in the colouring of synthetic fibres. These may be considered as being of four main types: the polyamides, polyesters, polyacrylics and polyolefins.

POLYAMIDES

The polyamides have a close-packed structure strongly held together by Van der Waals forces and hydrogen bonds, with three types of terminal group on the ends of the polymer chains.

Types of terminal group on polyamide chains

(amide group, carboxyl group, amino group)

Two types of dyestuff have been used successfully on this type of fibre, disperse dyes and anionic dyes. Disperse dyes, which include derivatives of diphenylamine and anthraquinone, and certain diazo dyes, are absorbed on to the surface of polyamide fibres and are also able to diffuse into the fibre pores. This produces even (level) dyeing but the colour is only moderately water-fast.

Increased fastness has been obtained by coordinating the dye molecules around a metal atom (pre-metallized disperse dyes) to increase the molecular size. Reactive disperse dyes have also been introduced ('Procinyl' dyes) which are capable of reacting with the amino and amide groups of the polyamide chain. Dyes of this type have been extensively used for the dyeing of nylon stockings.

Testing dyed nylon carpet fabric for light-fastness using a xenon arc lamp to produce simulated sunlight

Experimental warp knitted nylon fabrics being scoured before dyeing in the Pontypool laboratories of ICI Fibres Ltd

The anionic dyes, such as the acid azo wool dyes, triphenyl methane and chrome dyes, are attracted by the positively charged amino groups of the polyamide chain and then held in place by Van der Waals forces. The wet-fastness of acid dyes on polyamide fibres can be increased by after-treatment with a solution of tannic acid and potassium antimony tartrate.

POLYESTERS

The alignment of the polymer chain in polyesters such as 'Terylene' is so complete that an exceptionally close-packed structure results. Under these conditions the intermolecular bonding forces are very strong and the pore dimensions are correspondingly small. To make matters worse the polyester chain is markedly hydrophobic and possesses no active chemical groups.

In order to dye polyesters, therefore, it is necessary to increase the pore diameter of the fibres temporarily in order to allow penetration by the dyestuff. The pores can then be allowed to shrink to their original size, thus trapping the dye within the fibres. This enlargement of the fibre pores can either be carried out using a 'carrier' or by heating.

The carriers used for this purpose are usually short chain hydrocarbon or chlorinated hydrocarbon liquids, phenols or aromatic esters. These act like molecular wedges, forcing their way between the polymer fibres and opening them up to allow the larger dye molecules to enter. Complete removal of the carrier after dyeing is desirable otherwise the light fastness of the colour may be reduced, and also the smell of the carrier will be recognizable on the dyed material.

Heating to about 120°C causes a loosening of the forces binding the polymer chains as a result of their absorption of thermal energy. Also at this temperature the energy of the dye molecules increases and facilitates their penetration of the enlarged fibre pores. High temperature dyeing of this kind is carried out in stainless steel pressure vessels and is extensively used for processing 'Terylene'.

In 1949 the 'Thermosol' process was introduced by Du Pont for the dyeing of 'Dacron', the American equivalent of 'Terylene'. This involves heating the material to a temperature of between 180–220°C

Dyestuffs

whilst applying the dye (pad-dry heat technique). Under these conditions penetration and fixation are very rapid, the dyeing time often being reduced to a minute or two.

As with polyamides, disperse azo dyes are usually used for dyeing polyesters, but the high temperatures used in the pressurized and pad-dry techniques often cause dyes of this type to sublime. This effect can be reduced by the attachment of β-hydroxyethyl groups to amino nitrogen atoms attached to the dye molecule.

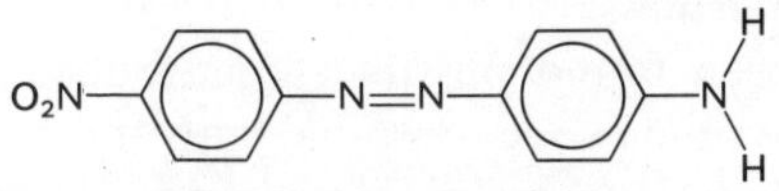

$$O_2N-\langle\bigcirc\rangle-N=N-\langle\bigcirc\rangle-N\begin{matrix}H\\ \\H\end{matrix}$$

poor sublimation fastness

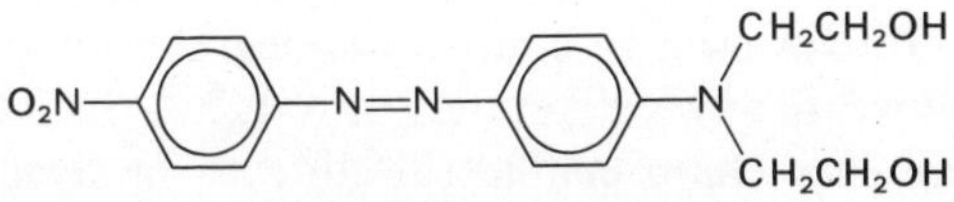

$$O_2N-\langle\bigcirc\rangle-N=N-\langle\bigcirc\rangle-N\begin{matrix}CH_2CH_2OH\\ \\CH_2CH_2OH\end{matrix}$$

good sublimation fastness

POLYACRYLICS

The acrylic polymer chains are linked into a hexagonal network produced as the result of hydrogen bonding between the nitrile (—CN) groups and hydrogen atoms on adjacent chains.

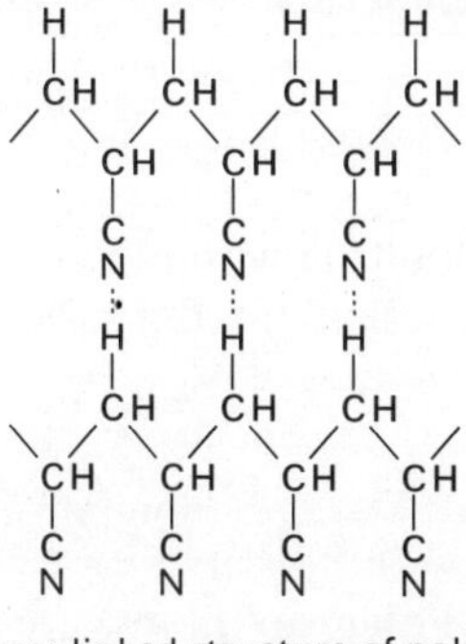

hexagonal cross linked structure of polyacrylonitrile

The low density of acrylics (1·17) suggests the probability of a rather more open structure than is encountered with the polyamides and polyesters. This has been found to be the case and permits greater ease of penetration by dye particles. The pore size of the acrylics has been further enlarged by copolymerization with a monomer which prevents the orderly packing of the copolymer molecules. This may be compared to the effect of parking lines of small cars with the inclusion occasionally of a bus which would disturb the orderly arrangement and leave unused spaces. Under these conditions disperse dyes can be used to produce pastel shades with good light-fastness.

It has also been found possible to introduce groups into the polyacrylonitrile framework to enable both anionic (acid type) and cationic (basic type) dyes to be used. Originally acrylic fibres often contained anionic sulphonic acid ($-SO_3^-$) residues which originated from additives used to control the degree of polymerization, but such groups are now deliberately introduced. Such anionic sites along the polymer chain enable the fibre to be dyed with basic dyestuffs.

Similarly by copolymerizing acrylonitrile with a basic monomer such as vinyl pyridine, cationic sites can be distributed along the polymer chain permitting the uptake of acidic dyes.

(a)
```
        \CH₂
   NC—CH
       \SO₃⁻  ⁺N—[ basic dye ]
       /
     CH₂
       \
      HC—CN
       /
```

(b)
```
        \
        HC—CN
        /
       CH₂         H
        \         N⁺
        HC—[ ⬡ ]   ⁻SO₃—[ acid dye ]
        /
       CH₂
        \
       NC—CH
        /
```

Copolymers of acrylonitrile with other vinyl derivatives such as 'Dynel' 'Teklan' and 'Creslan' (mod-acrylics) can similarly be dyed with disperse and basic dyes.

POLYOLEFINS

Polypropylene fibres ('Polyprop' fibres) are the only commercially important polyolefins at the time of writing. Their powerful hydrophobic nature and the complete absence of any polar groupings in their hydrocarbon framework have created an almost insuperable problem for the dyer. Disperse dyes produce only faint colouring, which is easily removed by rubbing or dry cleaning.

Some success has been achieved in incorporating a small percentage (0·1 %) of a metal salt in the fibre and then forming complexes between the metal and specially developed mordant-type dyes. Chromium, cobalt and nickel salts have been used for this purpose.

Other methods rely upon modifying the polypropylene by grafting active monomers on to the polymer backbone, or co-extruding molten polypropylene with another polymer. Attempts have also been made to mix coloured pigments with the molten polypropylene before extrusion.

POLYURETHANES

The polyurethane fibres represent the small but rapidly growing class of elastomeric fibres which have rubber-like properties. These are mostly used undyed, but pale shades with low fastness for light and water are produced using acid wool dyes and disperse dyes at low dyebath temperatures.

ORGANIC PIGMENTS

In addition to dyestuffs certain coloured organic substances are used as pigments. Unlike dyes, pigments are insoluble in the medium in which they are used and do not resist removal from surfaces to which they have been applied. It is usually necessary for a pigment to be highly opaque and light-fast and in certain cases to possess other properties such as resistance to heat, solvents and bleaches.

As in the case of dyes the azo pigments are of prime importance. The Hansa Yellows are prepared by the addition of a diazotized amine to a suspension of a finely powdered acetoacetarylamide in water. Thus Hansa Yellow G is prepared from diazotized m-nitro-p-toluidine and acetoacetanilide.

Yellow pigments of this group are fast to light, bright in colour and resistant to lime and oil solvents. They are widely used in the preparation of paints, oil-bound distempers and printing inks. Other azo pigments are produced by coupling diazotized aromatic

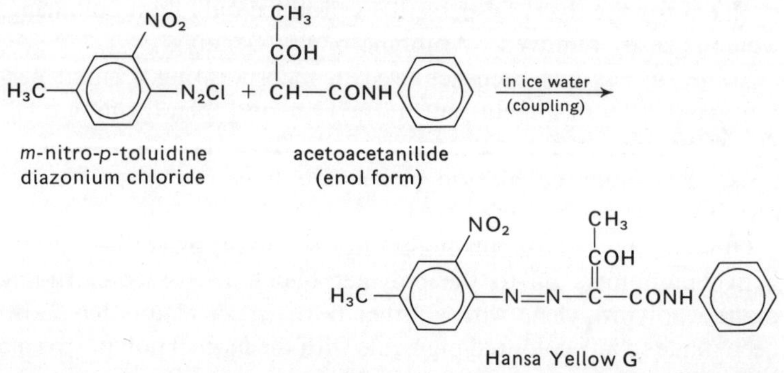

Fuels, Explosives and Dyestuffs

240

[Chemical reaction scheme:]

m-nitro-p-toluidine diazonium chloride (H₃C-C₆H₃(NO₂)-N₂Cl) + acetoacetanilide (enol form) (CH₃-C(OH)=CH-CONH-C₆H₅) → in ice water (coupling) →

Hansa Yellow G: H₃C-C₆H₃(NO₂)-N=N-C(CH₃)(OH)... wait, C(=COH with CH₃)-CONH-C₆H₅

amine and naphthylamine derivatives with β-naphthol (Toluidine Red).

The copper phthalocyanines giving the series of Monastral Fast Blues have already been mentioned and are in great demand for printing inks, plastics, paints and lacquers, and for the colouring of rubber and paper. Green phthalocyanines are produced by careful chlorination of the blue copper phthalocyanines.

Insoluble pigments used for printing inks are often prepared by precipitation from basic dyes using tannic acid solutions or solutions of phosphotungstates and phosphomolybdates. Metallic hydroxides have also been used to precipitate 'madder lakes' from alizarin.

A most unusual pigment which is fast to light and suitable for use in distempers and emulsion paints is Pigment Green B. This is produced by adding a solution of ferrous sulphate to the bisulphite addition compound of 1-nitroso-2-naphthol and then adding a saturated solution of sodium carbonate to precipitate the pigment.

Index

Acetaldehyde, 113, 123, 127
Acetic acid, 115, 123, 130, 149
Acetone, 112, 123, 126, 151, 152
Acetylene, 110, 112, 118, 132ff
Acid dyes, 216f
Acid magenta, 216
Acid milling dyes, 216
Acrolein, 115, 148
Acrylic acid, 134, 149
Acrylonitrile, 115, 133, 136, 149
Activated sludge, 101
Adipic acid, 158
Afterchrome mordanting, 215
Air pollution, 22ff
Akremite, 174
Alizarin, 193f, 207, 214
Allyl alcohol, 115, 148
Allyl chloride, 145, 147
Aluminium triethyl, 144
Amatol, 176
Ammonia, 135
 liquid, 118
Ammoniacal liquor, 33, 110
Ammonium nitrate, 173
Ammonium picrate, 177
Andrussow process, 136
ANFO, 174
Aniline black, 227
Aniline purple, 9
Aniline yellow, 196
Anthracene, 109
Anthracene oil, 108
Antifreeze, 143
Anti-knock, 84
Anti-oxidants, 84
Aquaflex, 186
Aquaspex, 186
Ardeer, 164
Aromatics, 157
Auramine, 218
Auxochromes, 206

Avaro process, 66
Aviation fuel, 46, 86
Azo Geranine, 216
Azoic dyes, 225

Bacon, Roger, 161
Bacton terminal, 91, 101
Bailer, 48
Basic dyes, 217
Bathochromic effect, 209
Bazooka, 187
Beaver Report, 22
Benfield plant, 43
Benzene, 49, 106
Benzene diazonium chloride, 196
Benzole, 104, 109
Bergius process, 46
Bethell, 103
Biazzi process, 168
Bickford, William, 162
Biodegradable detergents, 138
Birch Coppice, 29
Bismarck Brown, 196, 219
Bitumen, 56
Bituminous paint, 71
Black Creek, 47
Black powder, 161f, 171
Blasting gelatine, 166
Blasting oil, 164
Bleu de Lyons, 194, 216
Blown bitumen, 71
Bobbinite, 182
Bohn, René, 201
Boiling bed, 45
Bolt guns, 191
Boric acid, 116
Bottogas, 53
Bottoms, 53
BP acetic acid process, 130
Brenthol AS, 226
Briquettes, 27f

British Celanese, 123
British Coalite Co., 25
British Tar Co., 103
Bubble cap, 51f
Butadiene, 116, 156
Butagas, 84
Butanols, 154
Butylenes, 154
Butyl rubber, 155
Butyraldehyde, 135

Cable tool, 48
Calcium carbide, 118, 123
Caledon dyes, 212
Caledon Yellow, 202
Calor gas, 84
Calorific value, 19
Candida yeast, 80
Canvey terminal, 90
Caprolactam, 158
Carbojet, 88
Carbon black, 74, 136
Carbon tetrachloride, 137
Carburetting, 40
Carleton Ellis, 112
Carmoisine, 197
Catalytic rich gas, 81
Celanese Corporation, 115
Celite catalyst, 141
Cellition dyes, 228
Cellulose acetate, 123
CERCHAR, 88
Cetane number, 20
Channel black process, 136
Char, 28
Charcoal, 21
Chemical rocket engines, 86
Chemical Valley, 47
Chlorazol Brown, 219
Chlorazol dyes, 213
Chlorazol Orange, 220
Chloroethane, 137
Chloroprene, 133
Chromogens, 206
Chromophores, 206
Chrysoidine G, 196
Ciba Blue, 201
Cibacet dyes, 228
Cibacron dyes, 202, 231
Cigarette lighter fuel, 53

Clayton, John, 30
Clean Air Act, 5
Cleanglow, 27
Coal—
  ash content, 13f
  carbonization, 31
  chemical composition, 6, 14ff
  classification, 12
  coking properties, 20
  gas, 31ff
  Mines Regulations Act, 182
  production, 21
  source of chemicals, 7
Coalite, 25
Cobb, 41
Cochineal, 193, 214
Coke—
  metallurgical, 20, 34
  ovens, 34, 35
Colour index, 213, 226
Committee on Coal Derivatives, 43
Complete gasification, 38
Compound explosive, 179
Congo Red, 197, 219
Conjugated double bonds, 210
Continental Shelf Act, 91
Continuous gas-making processes, 77
Conventional gas manufacture, 40
Coomassie dyes, 217
Conversion processes, 57
Cordite, 171
Cordtex, 163
Cracking—
  catalytic, 60ff
  radiation, 63f
  thermal, 58f
Creosote, 109
Cross-conjugation, 210
Crystal Violet, 217
CTF (coal tar fuels), 107f
Cumene, 151
Cumene hydroperoxide, 152
Cuprofix Navy Blue, 220
Curme, George, 112
Cut-back bitumen, 72
Cyanine dyes, 205
Cyanuric chloride, 202ff
Cyclic gas making processes, 75ff
Cyclohexane, 157
Cyclohexanone, 157

Index 243

Cyclones, 62
Cyclonite, 180

Darby, Abraham, 103
DCL acetic acid process, 130
DDNP, 181
Denespex, 184
De Rerum Natura, 192
Desulphurizing, 70
Detergents, 152
Detonating cap, 164
Developed dyeing, 199
Dewaxing, 70
Diazamine Blue, 220
Diazo Brown, 219
Diazo dyes, 195
Dichroism, 211
Diesel fuel, 54, 85
Diethyl ketone, 114
Diethyl sulphate, 126, 141
Dimethyl formamide, 118
Dinitrobenzene, 175
Dinitronaphthalene, 177
Dinol, 181
Diphenylol propane, 148
Direct dyes, 198, 219
Disperse dyes, 228
Dispersol dyes, 228
Distillation—
 primary, 50
 secondary, 50
 vacuum, 54
Distillers Co. Ltd, 115, 123, 127
Dixon of Cookfield, 103
Doctor process, 69
Dodecane, 152
Double gas plants, 40
Drake, 'Colonel', 47
Drilling rigs, 93
Drimarene dyes, 231
Dubbs plant, 58
Dundonald, Earl, 103
Duo-Sol process, 57
Dye carriers, 229, 236
Dye classification, 212ff
Dyeing synthetic fibres, 232
Dyers Guilds, 193
Dyers thistle, 192
Dyestuffs, organic, 10, 192ff
Dynamite, 165

Edeleanu process, 56
Electrolinking, 38
Electrostatic precipitation, 32
Energy—
 requirements, 1
 transportation, 1
Epichlorhydrin, 147
Epoxy resin, 147
Eq.S. explosives, 184
Eriochrome Black T, 215
Ethane, 137
Ethanol, 141ff
Ethanolamines, 143
Ethyl benzene, 143
Ethyl chloride, 138
Ethylene, 112, 119ff
Ethylene chlorhydrin, 113, 142
Ethylene derivatives, 138
Ethylene dibromide, 141
Ethylene dichloride, 113, 140
Ethylene glycol, 113
Ethylene grids, 122
Ethylene oxide, 113, 142
Evelyn, John, 22
Explosion cladding, 190
Explosives—
 commercial, 7
 detonating, 8
 primary, 9
 propellants, 8
Extra dynamite, 166

Fast Blue Developer, 220
Fast Blue RR Base, 227
Fastness of dyes, 198
Fast Red AV, 197
Fast salts, 227
Fast Scarlet Salt, 227
Fibres, dyeing, 10
Filter drum, 70
Fireman, 182
Fischer–Tropsch, 43
Fixanol C, 220
Fluidized combustion, 21f
Fluidized sand cracking, 82
Fluorescers, 212
Formic acid, 115, 130
Fractionating, 51
Friedel–Craft reaction, 143
Fuchs, Sir Vivien, 186

Fuchsine, 217
Fuel oil, 85
Fuels—
  characteristics, 18ff
  impulse, 86
  origin, 12
  pattern of use, 1ff, 11
  smokeless, 25ff
  solid, 21ff
Fulvene, 210
Furnace black, 136

Gaine, 181
Gas Buggy operation, 99
Gas-fading, 229
Gas making—
  from coal, 30ff
  from liquid fuels, 72ff
Gas-oil, 54
Gasolines, 84
Gelignite, 166
Geophexes, 186
Giant Powder Co., 164
G.I. plant, 40
Girbotol process, 69
Gloco, 25
Glycerol, 147ff
Glyceryl trinitrate, 163, 171
Green oil, 108
Griess, Peter, 10, 195
Guncotton, 166
Gunpowder, 161
  blasting agent, 162

Haber process, 134
Hansa Yellow, 239
Hassi R'Mel, 90
Heavy oil, 107
Hell Gate reefs, 174
Hewett field, 96
Hexamethylene diamine, 156
Hexatriene, 210
Hexanitrodiphenylamine, 177
High temperature carbonization, 34
Hillman, 165
Hodsmen, 41
Hoechst process, 126
HMX, 181
Homefire, 29
Hortonspheres, 53

Hot acid process, 68
Huls process, 118
Hydrazine, 86
Hydrobel, 186
Hydrocarbon oxidation, 130
Hydrocracking, 117
Hydrogenation of coal, 46
Hydrogen cyanide, 136
Hydrosulphurization, 67, 69
Hyperconjugation, 211

ICI naphtha reforming process, 78ff
Igniter coal, 163
Ignition control additives, 84
Imbibition, 233
Indanthrene blue, 201
Indanthrone, 201
Indefatigable field, 96
Indigo, 192, 200
Indigosols, 202, 223
Indigotin, 199ff, 221ff
Indigo white, 222
Ingrain dyeing, 199
Inhibitors, 144
International Synthetic Rubber Co., 156
Isatin, 200
Isomerization, 67
Isoprene, 153
Isotactic, 145

Jetex, 178
Jet fuels, 86
Jones plant, 74

Katanol, 218
Kerosine, 53, 85
Ketene, 126
Kieselguhr, 165
Knocking, 84
Koppers–Totzec plant, 45
Kryptocyanine, 205f

Lacq field, 99
Lakes, 214
Laser gasification, 45
Lead azide, 179f
Lead picrate, 177
Lead styphnate, 178
Lean gas, 40
Leman bank, 96

Index

Leuco form of indigo, 200
Light distillate, 53
Light naphtha, 53
Light oil, 106
Lignite briquettes, 41
Liquefaction of coal, 46
Liquid hydrocarbons, 83, 117
Liquor trinitrini, 171
Lissamine dyes, 213, 217
Livesey extractor, 32
Logwood, 193
Low freezing explosives, 166
Low Temperature Carbonization Co., 25
Lox explosives, 174
LPG, 53, 84
Lube oils, 70
Lubricating oils, 70
Lurgi process, 41ff
Lyddite, 177

Madder, 193
Magenta, 194, 217
Malachite Green, 219
Mannitol hexanitrate, 172
Mauve, 193
Mauveine, 217
Mean pore size, 232
MEK, 155
Melanine pigments, 207
Melco, 112
Mercury fulminate, 164, 179
Metachrome process, 215
Metallic soaps, 71
Methane—
  carriers, 90
  derivatives, 134
  drainage, 101
  grid, 90ff
Methyl acrylate, 134
Methyl chloride, 136
Methyl dichloride, 136
Methylene blue, 219
Methylene dichloride, 137
Methyl nitrate, 171
Methyl violet, 194
Middle distillate, 53
Middle oil, 106
Midland-Yorkshire Tar Distillers Co., 105
Mod-acrylics, 238

Mogden plant, 102
Monastral Fast Blue, 205
Monex W rocket fuel, 190
Monobel, 173
Monoethyl sulphate, 125, 141
Mordant dyes, 198, 213ff
Mordants, 193
Multifilm washers, 25
Multiheat, 28
Murdoch, William, 30
Murex, 192

Naphtha, 53ff, 117
Naphthalene, 109
Naphthenates, 158
β-Naphthol, 199
Naphthols, 226
National Benzole, 104
Natural gas, 90ff
  composition, 18
  conversion, 97ff
  dehydration, 100
  design of burners, 19
  fields, 17f, 90, 99
  for synthesis, 117
  protein from, 89
  storage, 90f
Neolan dyes, 215
Neoprene, 133
Neutral Red, 219
Newman Spinney, 36
Nitrile rubber, 156
Nitrobenzene, 175
Nitroethane, 137
Nitroglycerine, 163
Nitromethane, 137
Nitro starch, 172
Nobel, Alfred, 164
Non-catalytic rich gas, 73
Nuclear power, 2ff

Octane rating, 20
Oil—
  deposits in UK, 16f
  origin, 15
  refining, 49ff
  seepage, 16
  types of crude, 16
Olefin gases, 119
Onia-Gegi plant, 76

Oppanol, 155
Orange, 11
Organic pigments, 239
Orientation of fibres, 232
Orthoflow process, 60
Otto process, 33
Oxidation dyes, 227f
Oxidation inhibitors, 71
Oxo process, 114
Oxyliquit, 174
β-Oxynaphthoic acid anilide, 199

P1 and P2 explosives, 182
P3 and P9 processes, 77
Padding, 199
Palatine dyes, 216
Paraffin wax, 117
Para Red, 225
Para Rosaniline, 218
Peat, 21
Pentaerythritol tetranitrate, 171
Pentolites, 172
Pentregel, 184
Perkin, William, 9, 104, 193ff
Permitted explosives, 182
Petrolatum, 56
Petroleum—
  chemical composition, 6
  chemical derivatives, 131ff
  crudes, 48ff
  ether, 53
  jelly, 56
  production, 47ff
  source of chemicals, 7, 110ff, 122ff
Phenol, 152
Phenyl-α-naphthylamine, 71
Phenyl glycine, 200
Phillips catalyst, 140
Phthalic anhydride, 159, 204
Phthalimide, 304
Phthalocyanins, 205
Phurnacite, 27
PIAT anti-tank weapon, 187
Picolines, 109
Picric acid, 177, 217
Pi electrons, 209
Pigment Green B, 240
Pipe still, 50
Pitch, 108
Platforming, 66

Pliny, 192
Polaris missile, 87
Polyacrylics, 237
Polyamides, 234
Polybutadiene, 156
Polycarbocyanines, 205
Polyesters, 236
Polyethylene, 138
  high density, 140
Polygenetic dyes, 214
Polymerization, 68
Polyolefins, 238
Polypropylene, 145
Polyurethanes, 239
Porphyrins, 204
Porta, Baptista, 161
Pot still, 106
Pre-metallized disperse dyes, 234
Prill, 174
Primuline, 199
ProAbd still, 106
Procinyl dyes, 232
Procion dyes, 202, 213, 231
Procion resins, 232
Producer gas, 40
Propagas, 84
Propane, 137
Propionaldehyde, 114
Propionic acid, 115, 130
iso-Propyl alcohol, 112, 151
Propylene, 112, 119, 145ff
Propylene dichloride, 145, 147
Propylene glycol, 150
iso-Propyl hydrogen sulphate, 151
Protein from gas-oil, 89
Pullars of Perth, 193
Pulsed infusion shotfiring, 184
Purifiers for coal gas, 33
PVC, 132
Pyridine, 109
Pyroglycerine, 163
Pyroligneous acid, 123

Quench oil, 65
Quinoline blue, 205
Quinonoid dyes, 202
Quinonoid structure, 210

Rackarock, 174
Raffinate, 56

Index

Ramsay, Sir William, 35
Raney nickel, 151
Rapidogen, 227
Raschig still, 106
Rayolanda process, 216
RDX, 181
Reactive dyes, 204, 230ff
Reactone dyes, 231
Reboiler, 53
Refinery gas, 117
Reforming—
   catalytic, 66f
   double thermal, 66
   thermal, 64ff
Remazol dyes, 231
Residue, 53
Resonance hybrids, 209
Rexco, 25ff
Rexcopine, 25
Rexcote, 25
Rhodamine Red, 219
Rich gas, 72
Risers, 50
Road pitch, 47
Robodorant A, 100
Roofing felt, 71
Rotary drilling, 48
Rubber dynamite, 166
Rubber solution, 53
Ruhrgas, 45

Sachsse process, 118
Safety fuse, 162
Safflower, 192
Saran, 132
Sasolburg, 43
SBP fuels, 54
Scavengers, 84
Schmid process, 167
Schonbein, 163
Scientific Design Co., 153
Scrubbing unit, 33
Sea Gem, 96
Segas plant, 75
Seismic surveying, 187
Selecto process, 57
Semi-coke, 28
Sensitol Red, 205
Setacyl dyes, 228
Sewage sludge, 101

Shaped explosives, 187
Shell 405 catalyst, 88
Shell phosphate process, 69
Shell synthesis gas process, 81
Siemens, Sir William, 36
Sigma electrons, 209
Slagging producers, 45
Sludge gas, 102
Smoke control zones, 23
Smokeless fuels—*see* Fuels
Sobrero, 163
Sodium dithionate, 200
Sodium perborate, 223
Sodium percarbonate, 223
Soledon dyes, 224
Solid fuel rockets, 86
Solihull process, 82
Solubilized disperse dyes, 229
Solvent extraction—
   coal, 47
   petroleum, 56
Spent oxide, 33
Sprengel explosives, 174
SSC plant, 77
Starter cartridges, 190
Steam cracking, 117, 122
Stockholm tar, 103
Stoke Orchard, 28
Straight run spirit, 53
Styrene, 144
Styrene butadiene rubber, 156
Sublimation fastness, 237
Substantive dyes, 219
Succinic acid, 115
Sulfolane, 157
Sulphur blacks, 225
Sulphur dyes, 224
Sweetening, 69
Swelling agents, 229
Synchromate process, 215
Synthesis gas, 38, 113
Synthol, 113

Tail gas, 53
Taninol, 218
Tar acids, 109
Tar bases, 109
Tardin, 30
Tar distillation, 105ff
Tartrazine, 216

Teepol, 58
T.E.L., 58
Terephthalic acid, 152, 159
p-Tertbutyl catechol, 144
Terylene, 143
Test galleries, 182
Tetan, 177
Tetrachloroethane, 133
Tetrachloroindigotin, 201
Tetrahydrothiophen, 79, 100
Tetralite, 179
Tetranitromethane, 177
Tetryl, 179
Texaco synthesis gas process, 81
Thermofor process, 60
Thermosol process, 81
Thiokol, 86
Thional dyes, 225
Thional Sky Blue, 225
Thylox process, 34
Titan 11 rocket, 86
Tollens, 171
Toluene, 158
Toluidine Red, 240
Torpex, 181
Trays, 50
Treating, 68
Trinitrobenzene, 175
Trinitrophenol, 177
Trinitroresorcinol, 178
Trinitrotoluene, 175f
Triphenylmethane dyes, 194
Triton, 175
Tula, 36
Tumescal D, 229
Turkey Red, 193, 214
Tyrian Purple, 192

UDMH, 86
Unconventional gas manufacture, 41

Underground gasification, 35
Unigel, 184

Van der Waals forces, 236
Vat dyes, 221
Velan NW, 232
Vertical retort—
  continuous, 31
  intermittent, 31
Visbreaking, 58
Viscosity index, 70
Viscostatic oils, 71
Vistanex, 155
Vinyl acetate, 113, 134
Vinyl acetylene, 133
Vinyl chloride, 132, 140
Vinylidene chloride, 132

Wacker process, 127, 142
Waltham Abbey, 162
Water gas, 40
Waxy raffinate, 70
Weaver units, 19
Wedge charges, 188
Westfield, 41
West Sole field, 96
White oils, 56
White spirit, 54
Williams, 47
Wilton still, 106
Winsor, Frederick, 31
Witt, 206
Woad, 192, 200
Wulff process, 118

Xylene, 106
Xylene Blue VS, 217
Xylenols, 106

Ziegler catalysts, 138, 156